Success at Statistics

A Worktext with Humor
Third Edition

Fred Pyrczak

California State University, Los Angeles

 Pyrczak Publishing
P.O. Box 250430 • Glendale, CA 91225

Project Director: Monica Lopez.

Editorial assistance provided by Sharon Young, Brenda Koplin, Kenneth Ornburn, Erica Simmons, Randall R. Bruce, and Cheryl Alcorn.

Cover design by Robert Kibler and Larry Nichols.

Cartoons by Randy Glasbergen. To view more of his cartoons, please visit www.glasbergen.com.

Printed in the United States of America by Malloy, Inc.

"Pyrczak Publishing" is an imprint of Fred Pyrczak, Publisher, A California Corporation.

ISBN 1-884585-53-1

Introduction

I wrote this book for students who need a solid grounding in basic statistical methods but who have some anxiety about taking a math-related class. Several features should help reduce anxiety:

1. I included numerous humorous, self-checking, riddle-based worksheets. Of course, I hope the humor will bring a bit of levity into an important course that sometimes seems to run dry. Much more important, however, is the instructional value of self-checking exercises that are riddle-based. If you have incorrect answers on a worksheet, the answer to the riddle will not make sense. Thus, you get immediate feedback that something is wrong without being given the correct answers. Reviewing the material usually will lead to correct answers in short order. Thus, you get to learn from your mistakes by being given a chance to reconsider your answers.

2. I have divided this book into short sections—each of which is much shorter than the typical chapter in a statistics textbook. This has several benefits. First, a short section of technical material—often only two or three pages long—is less intimidating than a long chapter. Even if you have difficulty with a section, knowing that it is short and contains only a limited amount of material makes it less intimidating as you reread it to achieve mastery.

    Second, your anxiety about mastering difficult material should be reduced as you experience success in learning the material. Mastering a section, even if it is only two pages long, helps reduce anxiety.

    Few things in learning are more frustrating than struggling through a long chapter of technical material and then, while attempting end-of-chapter exercises, realizing that the whole chapter needs to be reread and reviewed—an intimidating prospect. Frustration and intimidation often cause anxiety and failure.

Another advantage of short sections is that they allow instructors to customize their courses by (a) assigning only those sections needed to fulfill course objectives and (b) arranging the presentation of topics in a sequence best suited to their students' needs.

Where to Begin

If your basic math skills are rusty, I suggest that you begin with the Supplement in Part D of this book. It will help you review skills that you will use throughout this course.

Acknowledgments

Anne Hafner and Patricia Bates Simun, both of California State University, Los Angeles, and Roger A. Stewart of the University of Wyoming reviewed large portions of the First Edition of this book while it was in progress. Robert Morman of California State University, Los Angeles, reviewed the entire manuscript for the First Edition in its final stages and offered many useful suggestions. Richard Rasor of the American River College reviewed both the Second and Third Editions. In addition, Deborah M. Oh of California State University, Los Angeles, reviewed the Third Edition. To these colleagues, I am grateful for their invaluable assistance. Errors, of course, remain my responsibility.

About the Third Edition

In the Third Edition, you will find an entirely new part (Part C) titled "Putting It All Together." This part provides an overall review while showing how various statistical methods covered earlier in the book should be used in conjunction with each other. In this part, you will also find important new material on the limitations of signficance testing as well as a discussion of how statistical significance relates to practical significance.

On the humorous side, we commissioned Randy Glasbergen to prepare an additional 15 original cartoons for this edition—providing you with a total of 50 cartoons, which are interspersed throughout the book. If you enjoy these cartoons (as I'm sure you will), you can view more of his cartoons on his Web site at www.glasbergen.com.

Contacting the Author

You can provide me with feedback on this book by writing to me in care of info@pyrczak.com or by mailing a letter to the address shown on the title page of this book. Comments and suggestions that will lead to improvements in the next edition of the book are especially welcomed.

Fred Pyrczak
Glendale, California

"If we learn from our mistakes, shouldn't
I try to make as many mistakes as possible?"

"It's the new keyboard for the statistics lab. Once you learn how to use it, it will make computation of the standard deviation easier."

"Have you seen the latest longevity statistics for frogs who live in biology labs?"

Contents

PART B: Inferential Statistics

PART C: Putting It All Together

Section 1: Descriptive Versus Inferential Statistics

Descriptive statistics summarize **data**.[1] For example, suppose you have the scores on a standardized test for 500 subjects.[2] Instead of presenting a list of the 500 scores in a research report, you might present an *average* score, which describes the performance of the typical subject. You will learn about three different averages in this book, all of which are examples of descriptive statistics.

Note that a set of data does not always consist of scores. For instance, you might have data on the political affiliations of the residents of a community. To summarize these data, you might count how many are Democrats, Republicans, Independents, and so on, and then calculate the *percentages* of each. *A percentage* is a descriptive statistic that indicates how many units per 100 have a certain characteristic. Thus, if 42% of a group of people are Democrats, 42 out of each 100 people in the group are Democrats.

The summaries provided by descriptive statistics are usually much more concise than the original data set (e.g., an average is much more concise than a list of 500 scores). In addition, descriptive statistics help us interpret sets of data (e.g., an average helps us understand what is typical of a group).

Inferential statistics are tools that tell us how much confidence we can have when we generalize from a **sample** to a **population**.[3] You are familiar with national opinion polls in which a carefully drawn sample of only about 1,500 adults is used to estimate the opinions of the entire adult population of the United States. The pollster first calculates *descriptive statistics*, such as the percentage of respondents who are in favor of capital punishment and the percentage of respondents who are opposed. Having sampled, he or she knows that the results may not be accurate because the sample may not be representative. In fact, the pollster knows that there is a high probability that the results

---

[1]Note that the word *data* is plural.

[2]In psychology, sociology, and many related disciplines, the *subjects* of a study are called *participants* if they voluntarily agree to participate. *Respondents* and *examinees* are other terms that are commonly used when referring to individuals who are being studied.

[3]The word *inferential* is derived from the word *infer*. When we generalize from a sample to a population, we are *inferring* that the sample is representative of the population.

are off by at least a small amount. This is why pollsters often mention a *margin of error*, which is an inferential statistic. It is reported to warn that random sampling may produce errors, which should be considered when interpreting results.[4] For example, a weekly news magazine recently reported that in a national poll, 58% of the respondents believed that the economy was improving. A footnote indicated that the margin of error was ±1.8%. This means that the pollster was confident that the true percentage for the whole population was within 1.8 percentage points of 58%.[5]

A *population* is any group in which an investigator is interested. It may be large, such as all adults age 18 and over who reside in the United States, or it may be small, such as all registered nurses employed by a specific hospital.[6] An investigator is free to choose a population of interest and should clearly identify it when writing research reports. This helps consumers of research determine to whom the results apply.

A study in which all members of a population are included is called a *census*. A census is often feasible and desirable when working with small populations (e.g., an algebra teacher may want to pretest all students at the beginning of a course, which will help determine the level at which to begin instruction). Inferential statistics are *not* needed when describing the results of a census because there is no sampling error.

When a population is large, it is more economical to use only a sample of the population. With modern sampling techniques, highly accurate information can be obtained using relatively small samples. Various methods of sampling are described later in this book.

Descriptive tools such as averages and percentages for census data should be called *parameters*, not *statistics*. For example, an average score based on a study of a population (e.g., a census) should be referred to as a *parameter*, but

---

[4]The measurement techniques, especially the wording of the question(s), may also produce errors. That is why consumers of research usually want to know the exact wording of a survey question, especially if important decisions are to be made based on the results.

[5]Margins of error are described in detail in several sections in Part B of this book.

[6]Notice that all members of a population have at least one characteristic in common, such as all being registered nurses employed by a specific hospital.

an average for a sample should be referred to as a *statistic*. Here is a visual aid for remembering the difference:

$$\text{Statistics come from Samples}$$

$$\text{Parameters come from Populations}$$

Terms to Review Before Attempting Worksheet 1

descriptive statistics, data, average, percentage, inferential statistics, sample, population, margin of error, census, parameters

"My presentation is called *How to Overcome Your Fear of Statistics*. You can't. The End."

Worksheet 1: Descriptive Versus Inferential Statistics

> *Riddle*: What does the warning sign at the lake say?

DIRECTIONS: To find the answer to the riddle, write the answer to each question in the space immediately below it. The word in parentheses in the solution section next to the answer to the first question is the first word in the answer to the riddle, the word beside the answer to the second question is the second word, and so on.

1. Is an average a "descriptive" *or* an "inferential" statistic?

2. Do "descriptive" *or* "inferential" statistics help us generalize from a sample to a population?

3. What is the term for the entire group in which an investigator is interested?

4. Which statistic mentioned in Section 1 is an example of an inferential statistic?

5. When all members of a population in which an investigator is interested are included in a study, the study is called a_____.

6. "National opinion polls often use a sample of about 1,500 subjects." Is this statement true or false?

7. "Populations are always large." Is this statement true or false?

Worksheet 1 (Continued)

8. Is the term *data* "singular" *or* "plural"?

9. Should a percentage obtained from a census be referred to as a "statistic" *or* as a "parameter"?

10. Should an average for a sample be referred to as a *statistic*?

Solution section:

```
    scores (swim)   subjects (boat)   descriptive (never)   parameter (both)

       inferential (test)   information (and)   population (the)   census (of)

economical (bait)   singular (because)   pollster (fish)   plural (with)   false (water)

    true (the)   margin of error (depth)   investigator (warn)   yes (feet)   no (the)
```

Write the answer to the riddle here, putting one word on each line: _____ _____ _____ _____
_____ _____ _____ _____ _____
_____

Notes:

Section 2: Scales of Measurement

Scales of measurement (also known as *levels of measurement*) help investigators determine what types of statistical analyses are appropriate. It is important to master this section because it is referred to in a number of others that follow.

The lowest level of measurement is **nominal** (also known as *categorical*). It is helpful to think of this level as the *naming* level. Here are some examples:

→ Individuals name the political parties with which they are affiliated.

→ Individuals name their gender.

→ Individuals name the state in which they reside.

→ Individuals name the language they prefer to use at home.

Notice that the categories named by subjects in these examples do *not* put the subjects in any particular order. There is no basis on which we could all agree for saying that Republicans are logically higher or lower than Democrats. The same is true for gender, state of residence, and language preference.

The next level of measurement is **ordinal**. Ordinal measurement puts subjects in *order* from high to low, but it does *not* indicate how much higher or lower one subject is in relation to another. It is helpful to think of this level as the *ranking* level. To understand this level, consider these examples:

→ Individuals are ranked according to their height; with a rank of 1 for the tallest, a rank of 2 for the next tallest, and so on.

→ Three brands of hand lotion are ranked according to consumers' preferences for them, with a rank of 1 for the one most preferred, a rank of 2 for the next favorite, and so on.

→ Individuals rank the situation comedy programs on network television, giving a rank of 1 to their favorite, a rank of 2 to their next favorite, and so on.

In these examples, the measurements tell us the relative standings of individuals but not the amount of the differences among the individuals. For example, we know that an individual with a rank of 1 is taller than an individual with a rank of 2, but we do not know by how much. The first individual may be only one-quarter of an inch taller *or* two feet taller than the second.

The next two levels, ***interval*** and ***ratio***, both tell us by *how much* individuals differ. It is helpful to think of these as the *equal distance* levels. For example:

➔ The height of each individual measured to the nearest inch.

➔ The number of times each pigeon presses a button in the first minute after receiving a reward.

➔ The number of days each student is late arriving at school during the school year.

Notice that if one individual is 5'6" tall and another is 5'8" tall, we not only know the order of the individuals, but we also know by how many inches the individuals differ from each other. Both *interval* and *ratio* scales have equal intervals. For instance, the difference between 5'6" and 5'7" is the same as the difference between 5'7" and 5'8".

In most statistical analyses, *interval* and *ratio* measurements are analyzed in the same way. There is a mathematical difference, however. An *interval* scale does *not* have an absolute zero. For example, if we measure intelligence, we do not know exactly what constitutes zero intelligence and, thus, cannot measure it.[1] In contrast, a *ratio* scale has an absolute zero. For example, we know where the zero point is on a tape measure when we measure height.

If you are having trouble mastering levels of measurement, first memorize this environmentally friendly phrase:

$$\boxed{\textbf{N}\text{o } \textbf{O}\text{il } \textbf{I}\text{n } \textbf{R}\text{ivers}}$$

The first letters of the words (NOIR) are the first letters in the names of the four levels of measurement in order from the lowest to highest.

---

[1]Most applied researchers treat most sets of scores obtained by using standardized tests as being at the *interval* level, even though there is some controversy as to whether, for example, the difference between an IQ of 100 and 110 is the same as the difference between an IQ of 110 and 120.

Understanding scales of measurement is important in statistics because knowing them helps in the selection of appropriate statistics for a given set of data. For instance, you will learn about three averages in Sections 10 and 11. For a given set of data, typically only one average is selected to be included in a research report. The selection of the most appropriate average hinges, in large part, on the scale of measurement used to collect the data. In Section 11, you will learn more about how to use your knowledge of scales of measurement to select an average.

Terms to Review Before Attempting Worksheet 2

nominal, ordinal, interval, ratio

"Remember the old days when we used to eat his statistics homework?"

Worksheet 2: Scales of Measurement

> *Riddle*: According to Josh Jenkins, how do you know when you are making too many errors?

DIRECTIONS: For each example of measurement, circle the scale of measurement that it exemplifies. The word in parentheses to the right of the correct answer to the first question is the first word in the answer to the riddle, the word in parentheses to the right of the second correct answer is the second word, and so on.

1. Gender measured by asking each subject whether he or she is male or female.
   nominal (when)   ordinal (make)   interval (some)   ratio (only)

2. Weight measured in pounds and ounces using an accurate scale.
   nominal (only)   ordinal (if)   interval (of)   ratio (the)

3. Verbal aptitude measured by the College Board's Scholastic Aptitude Test: Verbal (assuming that each point from 200 to 800 represents an equal amount of aptitude).
   nominal (calculate)   ordinal (stumble)   interval (eraser)   ratio (become)

4. Cheerfulness measured by having a teacher give a rank of 1 to the student judged to be the most cheerful, a rank of 2 to the student judged to be the next most cheerful, etc.
   nominal (too)   ordinal (wears)   interval (for)   ratio (out)

5. The amount of juice in a bottle measured in fluid ounces.
   nominal (big)   ordinal (weak)   interval (inside)   ratio (out)

Worksheet 2 (Continued)

6. Length of a telephone conversation measured by recording the number of seconds from the beginning to the end of the conversation.

    nominal (count)   ordinal (person)   interval (mistakes)   ratio (ahead)

7. Height measured by having ten subjects take off their shoes and line up according to height with the tallest person at the front of the line and giving that person a score of 10, giving the next person a score of 9, etc.

    nominal (are)   ordinal (of)   interval (the)   ratio (made)

8. Writing skills measured by having all subjects write for 15 minutes on the same topic and having an English teacher put the essays in order from best to worst.

    nominal (when)   ordinal (the)   interval (person)   ratio (wrongly)

9. Birthplace measured by having each subject write the name of the country in which she or he was born.

    nominal (pencil)   ordinal (seeing)   interval (caught)   ratio (also)

Write the answer to the riddle here, putting one word on each line: _____ _____ _____ _____
_____ _____ _____ _____ _____

Notes:

Section 3: Frequencies, Percentages, and Proportions

A *frequency* is the number of individuals or cases; its symbol is f.[1] Another symbol, *N*, meaning *number of individuals*, is also used to stand for frequency.[2] Thus, if you see in a report that $f = 23$ for a score of 99, you know that 23 subjects had a score of 99.

A *percentage*, whose symbol is *P* or %, indicates the number per hundred who have a certain characteristic. Thus, if you are told that 44% of the individuals in a town are registered as Democrats, you know that for each 100 registered voters, 44 are Democrats. To determine how many (the *frequency*) are Democrats, multiply the total number of registered voters by .44. Thus, if there are 2,313 registered voters, .44 × 2,313 = 1,017.72 are Democrats. In a report in the general media, this would probably be rounded to 1,018. In a report in an academic journal, thesis, or dissertation, however, the answer is usually reported to two decimal places.

To calculate a percentage, use division. Consider this example: If 22 of 84 gifted children in a sample report being afraid of the dark, determine the percentage by dividing the number who are afraid by the total number of children and then multiply by 100; thus, 22 ÷ 84 = .2619 × 100 = 26.19%. This result means that, based on the sample, if you questioned *100 individuals* from the same population, you would expect about 26 of them to report being afraid of the dark. Notice that only 84 individuals were actually studied, yet the result is still based on 100.

A *proportion* is part of one (1). In the previous paragraph, the proportion of the sample of gifted children who are afraid of the dark is .2619 or .26—the answer obtained before multiplying by 100. This means that *twenty-six hundredths of the sample* is afraid of the dark. As you can see, proportions are harder to interpret than percentages. Thus, percentages are usually preferred to

---

[1] Note that f is italicized. If you do not have the ability to type in italics, underline the symbol. This applies to almost all statistical symbols. Also, pay attention to the case. A lower-case f stands for *frequency*; an upper-case *F* stands for another statistic described later in this book.

[2] An upper-case *N* should be used when describing a population; a lower-case *n* should be used when describing a sample.

proportions in all types of reporting. However, don't be surprised if you occasionally encounter proportions in scientific writing.

When reporting percentages, it is a good idea to also report the underlying frequencies because percentages alone can sometimes be misleading or not provide sufficient information. For example, if you read that 8% of the foreign language students at a university were majoring in Russian, you would not have enough information to make informed decisions on how to staff the foreign language department and how many classes in Russian to offer. If you read that $f = 12$ (8%), based on a total of 150 foreign language students, you would know that 12 students need to be accommodated.

Percentages are especially helpful when comparing two or more groups of different sizes. Consider these data:

	College A	College B
Total number of foreign language students	$f = 150$	$f = 350$
Russian majors	$f = 12$ (8%)	$f = 14$ (4%)

Notice that the frequencies tell us that College B has more Russian majors; the percentages tell us that *per 100 students*, College A has more Russian majors. Thus, if College A had 350 foreign language students, we would expect to find twice as many Russian majors at College A than at College B because 8% is twice 4%.

Note that if you are given the total sample size and the *percentage* who have a certain characteristic (e.g., 12% of the 1,000 voters in a town are registered as Independents), you can calculate the *number* who have the characteristic by multiplying the sample size by the proportion that corresponds to the percentage (e.g., $.12 \times 1,000 = 120$ voters who are registered as Independents). In other words, put a decimal point before the percentage (and remove the percentage sign), and multiply it by the total sample size.

Terms to Review Before Attempting Worksheet 3

frequency, percentage, proportion

Worksheet 3: Frequencies, Percentages, and Proportions

Riddle: How are famous surgeons similar to pelicans?

DIRECTIONS: To find the answer to the riddle, write the answer to each question in the space immediately below it. The word in parentheses in the solution section next to the answer to the first question is the first word in the answer to the riddle, the word beside the answer to the second question is the second word, and so on.

1. What is the symbol for frequency?

2. N is a symbol for what?

3. If 24% of the 1,511 students in a school qualify for a school lunch program, to two decimal places, how many qualify?

4. If 368 teenagers out of a population of 4,310 report that they have smoked at least one cigarette during the last year, to one decimal place, what percentage reported this?

5. If 3,100 people in a population of 46,785 complain of chronic headaches, to one decimal place, what percentage complain of this?

6. If 340 adults in a community of 6,532 report getting sufficient aerobic exercise, to three decimal places, what proportion of adults get this exercise?

Worksheet 3 (Continued)

7. To one decimal place, what is the percentage that corresponds to the proportion in the answer to question 6?

8. According to the text, are percentages or proportions usually easier for readers to comprehend?

9. According to the text, it is a good idea to report what statistic when reporting percentages?

10. If 6% of the students in School A and 6% of the students in School B are classified as gifted, do the two schools necessarily have the same number of students classified as gifted?

Solution section:

no (bills) frequencies (their) .066% (medicine) percentages (of)

5.2% (size) 9.5% (operate) .052 (the) 36.26 (hospital) 6.6% (by)

8.5% (recognized) 362.64 (be) yes (fly) number of individuals (can)

f (both) x (ocean) .85% (blood) 7.2% (fleeting) .66% (resting)

Write the answer to the riddle here, putting one word on each line: _____ _____ _____ _____

_____ _____ _____ _____ _____

_____

Section 4: Introduction to Frequency Distributions

A *frequency distribution* shows how many individuals had each score. Its purpose is to display scores so that they can be scanned by readers for an overview of the data. Table 4.1 shows the distribution of scores for a class of college students who took a 40-item basic math test. As you would expect, most students did quite well on this test. The frequency distribution makes it clear that most of the students marked 34 or more items correctly.

The distribution in Table 4.1 is *skewed*. In a skewed distribution, most of the scores are either near the top or bottom—with a scattering of scores toward the other end. Skewed distributions are discussed in more detail in later sections.

Table 4.1
Distribution of Basic Math Scores

X	f
37	8
36	4
35	3
34	6
33	1
32	3
31	1
30	1
29	2
28	2
27	0
26	0
25	1
24	1
$N =$	33

As you can see, a frequency distribution is an effective way to present data. There are, however, a number of conventions that are followed when constructing them. The most important are:

1. Each table should be given a number. Usually they are numbered sequentially, starting with the number 1. In the example on the previous page, 4 stands for the section number, and 1 indicates that it is the first table in that section.

2. Each table is given a brief descriptive *caption* (i.e., title). It is a good idea to name the population in the caption if frequency distributions are shown for two or more populations in the same report. Thus, the caption for Table 4.1 might be *Distribution of Basic Math Scores for College Students*.

3. The number and caption are placed at the top of the table.

4. There is a horizontal line that sets off the table number and caption from the body of the table. This line is called the *rule*. There is also a rule at the bottom of the table to indicate where it ends.

5. The symbol X stands for the scores. Be sure to use an upper-case italicized X because a lower-case italicized x has another meaning, which is discussed later in this book.

6. The scores are listed in order with the highest score obtained by an individual placed at the top, and the lowest score obtained by an individual placed at the bottom.

7. The symbol f stands for frequency or number of individuals. The symbol N may also be used at the top of the second column.

8. When no individuals have a score—such as scores 26 and 27—a frequency of zero is entered.

9. The sum of the frequencies is shown at the bottom. Instead of an N, it is acceptable to use ΣN, which means *sum of the number of cases,* or Σf, which means *sum of the frequencies*.

Terms to Review Before Attempting Worksheet 4

frequency distribution, skewed, caption, rule, X, f, N, Σ

Worksheet 4: Introduction to Frequency Distributions

> *Riddle*: Arthur Godfrey said that he was proud to be paying taxes. What else did he say?

DIRECTIONS: To find the answer to the riddle, write the answer to each question in the space immediately below it. The word in parentheses in the solution section next to the answer to the first question is the first word in the answer to the riddle, the word beside the answer to the second question is the second word, and so on.

1. According to Table 4.1, how many subjects had a score of 32?

2. What type of distribution has most of the scores either near the top or bottom with a scattering of scores toward the other end?

3. What is the formal name for the title of a table?

4. The horizontal lines that identify the top and bottom of a table are called what?

5. For what does *X* stand?

6. Should the highest or the lowest score be at the bottom of a frequency distribution?

Worksheet 4 (Continued)

7. Should the sum of the frequencies be shown in a frequency distribution?

8. What symbol may be substituted for f in a frequency distribution?

9. What is the symbol for the *sum of the frequencies*?

10. If a score in the middle of a distribution was obtained by none of the subjects, should the score be omitted from the list of scores in the first column of a frequency distribution?

Solution section:

> no (money) N (half) X (politics)   normal (president)   headline (going)   free (it)
>
> Σf (the)   2 (party)   0 (helpful)   3 (he)   skewed (could)   rules (just)
>
> caption (be)   score (as)   yes (for)   highest (courts)   double lines (it)
>
> lowest (proud) XF (views) Σx (audit)   omit (IRS)   unbalanced (was)

Write the answer to the riddle here, putting one word on each line: _____ _____ _____ _____
_____ _____ _____ _____ _____
_____

Section 5: Frequency Distribution for Grouped Data

Because the purpose of a frequency distribution is to organize data so that it can be presented concisely, do not list more than about 20 scores. If there are more than 20, you should group the scores as shown in Table 5.1.

Table 5.1
Frequency Distribution of Scores on
the AIDS Knowledge Test

X	tally marks	f
39–41	/	1
36–38	/	1
33–35		0
30–32	///	3
27–29	/////	5
24–26	/////	5
21–23	///// /	6
18–20	///// //	7
15–17	////	4
12–14		0
9–11	/	1
6–8	/	1
	Σf =	34

Table 5.1 represents the scores shown in Table 5.2. Notice that Table 5.1 is much more effective than Table 5.2 in communicating the performance of the

Table 5.2
Unarranged Scores on the AIDS Knowledge Test

30	10	15	17	41	37	23	29	18	6
16	17	21	21	23	29	29	21	22	31
32	27	27	24	18	18	25	26	18	25
26	19	20	19						

21

subjects on the test. It is clear from Table 5.1 that most subjects have scores between 15 and 32. This is not immediately obvious in Table 5.2.

These are some guidelines for constructing frequency distributions for grouped data:

1. There should be about 10 to 20 groups of scores. These groups are called *score intervals*. The bottom interval in Table 5.1 is for scores 6 through 8; this covers 3 points; thus, the *interval size* is 3 points.

2. All score intervals must have the same interval size. Note: Start by listing the bottom interval first; build up until you have included the highest score; make the top interval the same size as the others even if the top interval extends beyond the highest score earned.

3. To estimate the interval size, subtract the lowest score from the highest score (i.e., $41 - 6 = 35$) and divide the answer by 15 ($35/15 = 2.33$).[1] Thus, each interval should be *about* 2 points wide in order to yield 10 to 20 intervals; in this case, 3 was used.

4. Most people use an odd interval size because this makes it easier to plot a distribution on graph paper, which is described in later sections. Thus, in Table 5.1, an interval size of 3 was used instead of 2, which was obtained in the previous guideline by dividing by 15.

5. When you tally, work from the unarranged scores to the distribution. For example, put your finger on the first score (30) in Table 5.2. Make a tally to the right of 30 in Table 5.1. Then cross off the 30 in Table 5.2. Repeat the process for each of the remaining scores. Some students use the reverse procedure; they take the top score interval in the distribution (39–41) and then search through all the unarranged scores looking for scores of 39, 40, and 41. This is a slow method and is likely to lead to errors.

6. As a partial check on your work, make sure that the sum of the frequencies is the same as the total number of scores that you started with.

7. Erase the tally marks before submitting your distribution.

It is often helpful to your reader to also present percentages. In Table 5.3, a column showing the percentages has been added. To obtain the percentages,

---

[1]The constant 15 is used because it is halfway between 10 and 20. By always using 15, you will get the approximate interval size that will yield between 10 and 20 intervals.

divide each frequency (*f*) by the total number of cases (Σf), and then multiply by 100. Here are some examples:

1 divided by 34 = 0.0294 × 100 = 2.94 = 2.9% (the percentage for the bottom interval).

4 divided by 34 = 0.1176 × 100 = 11.76 = 11.8%.

Note that the percentages are calculated to two decimal places and rounded to one; it is customary to report them to one or two places in scientific reports. Also note that the percentages will not always sum to exactly 100% because of rounding. Don't be surprised if you get slightly more or less than 100%.

Table 5.3
Frequency Distribution of Scores on the AIDS Knowledge Test

X	tally marks	*f*	*P*
39–41	/	1	2.9
36–38	/	1	2.9
33–35		0	0.0
30–32	///	3	8.8
27–29	/////	5	14.7
24–26	/////	5	14.7
21–23	///// /	6	17.6
18–20	///// //	7	20.6
15–17	////	4	11.8
12–14		0	0.0
9–11	/	1	2.9
6–8	/	1	2.9
	$\Sigma f =$	34	99.8

Terms to Review Before Attempting Worksheet 5

score intervals, interval size

"According to a recent survey, 51% is a majority."

Worksheet 5: Frequency Distribution for Grouped Data

> **Riddle**: According to Sam Ewing, what will happen when the meek inherit the earth?

DIRECTIONS: To find the answer to the riddle, construct a frequency distribution for the scores in the box below. Follow all of the guidelines in Section 5 and include a column with percentages rounded to one decimal place.

Answer each question in the space immediately below it. The word in parentheses in the solution section next to the answer to the first question is the first word in the answer to the riddle, the word beside the answer to the second question is the second word, and so on.

> 15, 12, 18, 14, 36, 37, 42, 45, 48, 19, 32,
> 29, 18, 21, 36, 40, 41, 54, 33, 35, 34, 42,
> 40, 36, 33, 19, 15, 39

1. What value did you obtain when you subtracted the lowest score from the highest score?

2. Based on a divisor of 15, what odd number is an appropriate interval size?

3. What scores are shown in the first column for the bottom (lowest) score interval?

4. What scores are shown in the first column for the top (highest) score interval?

5. What is the frequency for the highest score interval?

25

Worksheet 5 (Continued)

6. What is the frequency for the 33–35 score interval?

7. What is the sum of the frequencies?

8. What is the percentage for the 51–53 score interval?

9. What is the percentage for the 42–44 score interval?

10. What is the sum of the percentages?

Solution section:

43 (religion)	2 (angels)	10–12 (only)	52–54 (civilization)	71% (all)
100.1% (details)	100.0% (crook)	0.0% (out)	99.9% (he)	7.1% (the)
42 (the)	3 (lawyers)	54–56 (be)	10.0% (being)	12–14 (will)	28 (work)
1 (there)	24–26 (fruitful)	4 (to)	55 (bible)	98.9% (Eden)

Write the answer to the riddle here, putting one word on each line: _____ _____ _____ _____
_____ _____ _____ _____ _____
_____

Section 6: Cumulative Frequencies, Cumulative Percentages, and Percentile Ranks

A *cumulative frequency* (*cf*) indicates how many individuals are *in and below* a given score interval. Study these examples, which are based on the data in Table 6.1:

In the bottom interval, 6–8, there is one case *in* the interval and zero cases *below* the interval. Therefore, the *cf* for this interval is $1 + 0 = 1$. (Remember that the frequency column, *f*, tells us the number of cases in each score interval.)

Table 6.1
Frequency Distribution of Scores on the AIDS Knowledge Test with Cumulative Frequencies and Cumulative Percentages

X	f	cf	%	cum %
39–41	1	34	2.9	99.8
36–38	1	33	2.9	96.9
33–35	0	32	0.0	94.0
30–32	3	32	8.8	94.0
27–29	5	29	14.7	85.2
24–26	5	24	14.7	70.5
21–23	6	19	17.6	55.8
18–20	7	13	20.6	38.2
15–17	4	6	11.8	17.6
12–14	0	2	0.0	5.8
9–11	1	2	2.9	5.8
6–8	1	1	2.9	2.9
	34		99.8	

In the next interval, 9–11, there is one case *in* the interval and one case *below* the interval. Therefore, the *cf* for this interval is 1 + 1 = 2.

In the next interval, 12–14, there are zero cases *in* the interval and a total of two cases in *all* the intervals below it. Therefore, 0 + 2 = 2.

The ***cumulative percentage*** (***cum %***) column has a similar meaning as the cumulative frequency column: Each entry indicates the *percentage in* and *below* a given score interval. For example:

For the 39–41 score interval, 99.8% scored *in* and *below* that interval. (Note that it is not 100.0% because of errors created by rounding to one decimal place.)

For the 36–38 score interval, 96.9% scored *in* and *below* that interval.

The cumulative percentages are approximate ***percentile ranks***, which indicate the percentage who scored at or below a given score level.[1] For example, we could report to subjects with scores of 27, 28, and 29 that their percentile rank is 85 (based on the *cum %* of 85.2)—indicating that, relative to the group, their scores are high because they scored as high or higher than 85% of the other subjects. Reporting percentile ranks is usually more informative than reporting ***raw scores*** (i.e., the number of points earned).

Terms to Review Before Attempting Worksheet 6

cumulative frequency, cumulative percentage,
percentile ranks, raw scores

---

[1]More precise methods are usually covered in measurement courses. This method is sufficient for most applied applications such as in classroom settings.

Worksheet 6: Cumulative Frequencies, Cumulative Percentages, and Percentile Ranks

> *Riddle*: What proves that many gas station owners think toilet paper is worth more than money?

DIRECTIONS: To find the answer to the riddle, construct columns for the cumulative frequencies, percentages, and cumulative percentages in Table 6.2 below.

Write the answer to each question in the space immediately below it. The word in parentheses in the solution section next to the answer to the first question is the first word in the answer to the riddle, the word beside the answer to the second question is the second word, and so on.

Table 6.2
Distribution for Worksheet 6

X	f
67–71	2
62–66	1
57–61	2
52–56	4
47–51	6
42–46	7
37–41	8
32–36	5
27–31	2
22–26	0
17–21	2
12–16	1
N =	40

Worksheet 6 (Continued)

1. What is the cumulative frequency for the 67–71 score interval?

2. What is the cumulative frequency for the 22–26 score interval?

3. How many subjects scored in and below the 12–16 score interval?

4. How many subjects scored in and below the 32–36 score interval?

5. What percentage of the subjects scored in the 22–26 score interval?

6. What is the cumulative percentage for the 22–26 score interval?

7. What is the cumulative percentage for the 57–61 score interval?

8. What is the cumulative percentage for the 67–71 score interval?

9. What is the approximate percentile rank for a score of 39?

10. What is the approximate percentile rank for a score of 64?

Worksheet 6 (Continued)

Solution section:

95 (unlocked)	45 (registers)	43 (press)	100.0 (cash)	2 (between)	4 (need)
10 (toilets)	0.0 (locked)	39 (wishes)	8.5 (only)	9.5 (tires)	7.5 (but)
40 (they)	1 (their)	99.9 (pump)	92.5 (their)	90.0 (being)	11 (gasoline)
10.5 (dollars)	15.5 (cars)	3 (keep)	69.5 (policy)	46 (windshield)	

Write the answer to the riddle here, putting one word on each line: _____ _____ _____ _____
_____ _____ _____ _____ _____
_____

GLASBERGEN

"I'm learning how to relax, doctor — but I want to relax better and faster! I WANT TO BE IN THE TOP PERCENTILE RANK OF RELAXATION!"

Notes:

Section 7: Histograms

A distribution of scores may be displayed in a statistical *figure* (i.e., a drawing or graph) known as a *histogram*, which is a vertical bar graph. Figure 7.1 shows a histogram for the distribution in Table 6.2 in Section 6. As you can see, the histogram is a graphic alternative for presenting the data in a frequency distribution.[1]

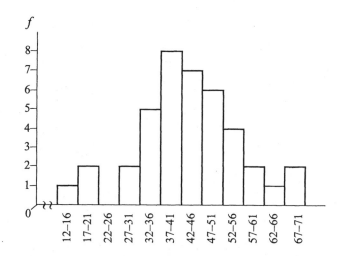

Figure 7.1. Histogram for data in Table 6.2 in Worksheet 6.

In Figure 7.1, frequencies are shown on the vertical axis (i.e., the *ordinate*), and the scores (or score intervals) are shown on the horizontal axis (i.e., the *abscissa*). A histogram may also be prepared for percentages or proportions by placing them instead of the frequencies on the vertical axis.

The following are guidelines for preparing histograms:
1. Draw the histogram on graph paper, if possible.
2. Plan ahead so that the vertical axis is somewhat shorter than the horizontal axis. A ratio of 3 to 5 is about right; that is, if the vertical axis is about 3

---

[1]The choice between a frequency distribution and a histogram depends on three factors: (1) an unsophisticated audience might find a histogram more comprehensible; (2) a histogram may be more visually appealing and more likely to draw readers' attention; and (3) histograms should be professionally drawn whenever possible, making them more difficult and expensive to produce.

inches, the horizontal axis should be about 5. This may mean that on the vertical axis, you may need to let each line on the graph paper represent more than one frequency. For example, if the highest frequency in a distribution is 100, you might need to let each line represent 10 cases in order to keep your figure reasonable in size.

3. Place the scores on the horizontal axis with the lowest score interval on the left and the highest on the right.[2]

4. Start the frequencies at the bottom of the vertical axis with zero, even if there is no score interval with a frequency of zero. Starting at a higher frequency can produce a truncated, misleading figure. Label the vertical axis with f if you are reporting frequencies (or with P if you are reporting percentages).

5. Label the histogram with a number and brief title (i.e., caption). Note that because the histogram is a statistical figure (i.e., a drawing or graph), it is called a *figure* and not a *table*.

6. Note that in many fields of study, *figure* numbers and titles are placed *below* the figure; *table* numbers and titles are placed *above* the table.

7. All bars must be made equal in width to avoid producing a misleading figure.

8. If there is a score interval within a distribution with a frequency of zero, a space should appear there; notice the space for the interval from 22–26, which has a frequency of zero.

Terms to Review Before Attempting Worksheet 7

figure, histogram, ordinate, abscissa

---

[2]Notice in Figure 7.1 that the scores between zero and 12 are not shown because no one had those scores. The horizontal line is broken just to the right of zero to indicate that some scores are not shown.

Worksheet 7: Histograms

Riddle: According to Benjamin Franklin, what should you do with your eyes to have a successful marriage?

DIRECTIONS: To find the answer to the riddle, write the answer to each question in the space immediately below it. The word in parentheses in the solution section next to the answer to the first question is the first word in the answer to the riddle, the word beside the answer to the second question is the second word, and so on.

1. "A histogram is an example of a statistical table." Is this statement true or false?

2. "A histogram is an example of a graph." Is this statement true or false?

3. What is the formal name of the vertical axis in a graph?

4. What is the formal name of the horizontal axis in a graph?

5. In a histogram, which axis (vertical or horizontal) should be longer?

6. What should be listed along the horizontal axis of a histogram?

Worksheet 7 (Continued)

7. What is the lowest frequency that should be shown on a histogram?

8. Should the lowest score (or score interval) in a histogram be on the left or the right side of the figure?

9. Should the bars in a histogram be of equal width even if the frequencies vary within the distribution?

10. In Figure 7.1, which score interval has the highest frequency?

11. In Figure 7.1, which score interval has the lowest frequency?

Solution section:

zero (marriage)	no (running)	false (keep)	37–41 (shut)	one (vows)
true (your)	27–31 (ring)	right (someone)	left (and)	yes (half)	abscissa (wide)
caption (feel)	horizontal (open)	scores or score intervals (before)	ordinate (eyes)	
vertical (love)	frequencies (begin)	22–26 (afterwards)	42–46 (he)	

Write the answer to the riddle here, putting one word on each line: _____ _____ _____ _____
_____ _____ _____ _____ _____
_____ _____

Section 8: Frequency Polygons

A *frequency polygon* represents a frequency distribution in graphic form; it is an alternative to a histogram for presenting a distribution. The polygon shown in Figure 8.1 is based on the distribution in Table 8.1.

Table 8.1
Distribution of Depression Scores

X	f
22	1
21	3
20	4
19	8
18	5
17	2
16	0
15	1
	N = 24

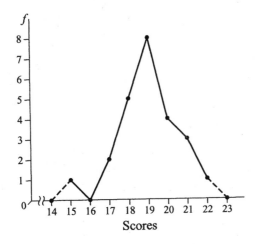

Figure 8.1. Frequency polygon based on Table 8.1.

37

These are guidelines for constructing frequency polygons:

1. If possible, draw it on graph paper.

2. List the frequencies (f) on the vertical axis (i.e., the ordinate or y-axis) and label the axis with an f.

3. List the scores on the horizontal axis (i.e., the abscissa or x-axis) and label the axis (in this case, "Depression Scores"). Begin with one score lower than any obtained by a subject and end with one score higher than any obtained; this will make the polygon "rest" or "anchor" on the horizontal axis. For example, in Table 8.1, the lowest score is 15, but the polygon begins with an anchor score of 14—at which point the line rests on the horizontal axis at a frequency of zero. At the upper end of the distribution, the polygon is anchored at a score of 23. A dashed line has been used to indicate that the anchor scores of 14 and 23 were not obtained by any of the examinees.

4. Plan ahead to make the horizontal axis somewhat longer than the vertical axis. A ratio of about 3 to 5 is about right (i.e., if the vertical axis is about 3 inches, the horizontal axis should be about 5). To make this happen, you may need to allow more than one space on your graph paper among the frequencies or the scores that you list.

5. Put a dot on the intersection on the graph paper where each score intersects with each frequency. For example, in Figure 8.1, a dot was placed where a score of 15 intersects a frequency of 1 (one). Then connect the dots with a ruler.

6. Don't forget to connect dots that are on the horizontal axis. For example, in Figure 8.1, there is a frequency of 0 (zero) at a score of 16; at that point, the line drops to the horizontal axis.

7. Label the polygon with a number and brief title (i.e., caption).

If the scores have first been grouped into score intervals, as in Table 8.2, and you wish to plot a frequency polygon, you must first determine the ***midpoint*** of each interval. These are shown in Table 8.2. For example, 40 is the midpoint (i.e., middle score) of the 39–41 score interval.[1] List the midpoints on the horizontal axis for your polygon, as shown in Figure 8.2. Also, list one

---

[1] By using an odd number as the size of the score interval, the midpoint will be a whole number.

midpoint higher and one midpoint lower than actually shown in the frequency distribution. This makes the polygon "rest" or "anchor" on the horizontal axis. Remember that anchor scores are not actual scores earned by anyone.

Table 8.2
Frequency Distribution of Number of Days Served in Prison

X	midpoints	f
39–41	40	1
36–38	37	1
33–35	34	0
30–32	31	4
27–29	28	4
24–26	25	4
21–23	22	6
18–20	19	6
15–17	16	4
12–14	13	0
9–11	10	1
6–8	7	1
	N =	32

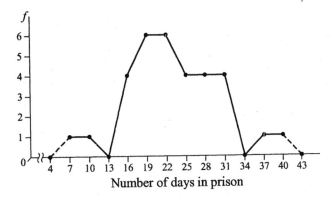

Figure 8.2. Frequency polygon based on Table 8.2.

With large samples, the lines on a frequency distribution usually will be fairly smooth—unlike our examples in which there are jagged peaks. When a smooth polygon emerges, it is usually referred to as a **curve**. Types of curves are described in the next section.

Terms to Review Before Attempting Worksheet 8

frequency polygon, midpoint, curve

"Professors are fighting back against students who cut class. Today I was the victim of a drive-by statistics quiz!"

Worksheet 8: Frequency Polygons

Riddle: Why should you never try to teach a pig
to sing?

DIRECTIONS: Construct a frequency polygon for the distribution shown below. Follow all the guidelines described in Section 8. Refer to your work while answering the questions.

To find the answer to the riddle, write the answer to each question in the space immediately below it. The word in parentheses in the solution section next to the answer to the first question is the first word in the answer to the riddle, the word beside the answer to the second question is the second word, and so on.

Table 8.3
*Frequency Distribution for
Worksheet 8*

X	f
95–99	3
90–94	4
85–89	6
80–84	8
75–79	11
70–74	12
65–69	15
60–64	10
55–59	5
50–54	4
45–49	0
40–44	2
N =	80

Worksheet 8 (Continued)

1. What is the midpoint for the 40–44 score interval?

2. At how many points does the polygon touch the horizontal axis (at a frequency of zero)?

3. How many midpoints are listed along the horizontal axis?

4. The highest point of the polygon is at which midpoint?

5. What is shown on the vertical axis?

6. What is the midpoint at the far right-hand end of the horizontal axis?

7. Is the vertical axis somewhat longer than the horizontal axis?

8. The highest dot is placed at what frequency?

9. Does the figure have a number and a title (caption)?

Worksheet 8 (Continued)

Solution section:

8 (farm)   1 (tune)   midpoints (song)   yes (pig)   10 (same)

97 (success)   19 (a)   3 (wastes)   42 (it)   37 (being)   8 (small)   23 (in)

frequencies (and)   2 (belief)   figure (helpless)   no (annoys)   table (seeing)

14 (your)   67 (time)   102 (it)   15 (the)   0.00 (barn)   99 (farmer)

Write the answer to the riddle here, putting one word on each line: _____   _____   _____   _____,
_____   _____   _____   _____   _____

Notes:

Section 9: Shapes of Distributions

After a set of scores has been organized from lowest to highest, it is referred to as a ***distribution***. One of the best ways to see a distribution's shape is to make a drawing of it with the scores on the horizontal axis and the frequencies (i.e., number of cases or subjects) on the vertical axis. The result is called a frequency polygon (see Section 8). When the number of cases is large, the polygon usually has a smooth shape, often referred to as a ***curve***.

The most important shape is that of the ***normal curve***—often called the bell-shaped curve—which is illustrated in Figure 9.1. It is important for two reasons. First, it is a shape often found in nature. For example, the heights of women in a large population are normally distributed. There are small numbers of very short women, which is why the curve is low on the left; many women of about average height, which is why the curve is high in the middle; and there are small numbers of very tall women. Here is another example: The distribution of average annual rainfall in Los Angeles over the past 105 years is approximately normal. There are a very small number of years in which there was extremely little rainfall, many years with about average rainfall, and a very small number of years with a great deal of rainfall. The second reason the normal curve is important is because it is used in inferential statistics, a topic that is covered in the second half of this book.

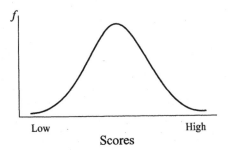

Figure 9.1. A normal distribution.

Some distributions are *skewed*. For example, if you plot the distribution of income for a large population, you will find that it has a ***positive skew*** (i.e., is skewed to the right). Examine Figure 9.2. It indicates that there are large numbers of people with relatively small incomes; thus, the curve is high on the left. The curve drops off dramatically to the right forming a tail on the right; this tail is created by the small numbers of individuals with very high incomes. Skewed distributions are named for their tails. On a number line, positive numbers are to the right; hence, the term *positive* skew.

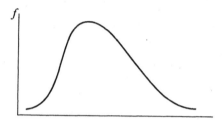

Figure 9.2. A distribution skewed to the right (positive skew).

When the tail is to the left, a distribution is said to have a ***negative skew*** (i.e., skewed to the left). See Figure 9.3. A negative skew would be formed if a large population of people were tested on skills in which they have been thoroughly trained. For example, if you tested a very large population of recent nursing school graduates on basic nursing skills, a distribution with a negative skew should emerge. There should be a large number of nurses with high scores, but there should also be a tail to the left created by a small number of nurses who, for one reason or another (such as being physically ill the day the test was administered), did not perform well on the test.

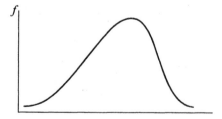

Figure 9.3. A distribution skewed to the left (negative skew).

Bimodal distributions are less frequently found. These have two high points. A curve such as that in Figure 9.4 is called bimodal even though the two high points are not equal in height. Such a curve is most likely to emerge when human intervention or some rare event has changed the composition of a population. An example of human intervention is the decision of a school board to establish a school to which only high achievers and low achievers are admitted because the school is to be a model of peer tutoring in which the high achievers tutor the low achievers. The distribution of scores on an achievement test for students admitted to the school should be bimodal. Another example is when a civil war in a country costs the lives of many young adults, the distribution of age after the war might be bimodal, with a dip in the middle.

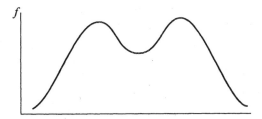

Figure 9.4. A bimodal distribution.

The shape of a distribution should always be examined before proceeding with additional statistical analyses. The shape has important implications for determining which average to compute—a topic that will be taken up next in this book—and for determining which other statistics are appropriate to compute.

Terms to Review Before Attempting Worksheet 9

distribution, curve, normal curve, skewed, positive skew, negative skew, bimodal distributions

Worksheet 9: Shapes of Distributions

Riddle: Woody Allen said, "It's not that I'm afraid to die, it's just that..."

DIRECTIONS: To find the answer to the riddle, write the answer to each question in the space immediately below it. The word in parentheses in the solution section next to the answer to the first question is the first word in the answer to the riddle, the word beside the answer to the second question is the second word, and so on.

1. What should be placed on the horizontal axis of a frequency polygon?

2. What is the most important type of curve?

3. A curve with a tail to the left is said to have what type of skew?

4. A curve with a tail to the right is said to have what type of skew?

5. "Income in a large population is usually normally distributed." Is this statement true or false?

6. In a normal curve, is the highest point to the left, to the right, or in the middle of the distribution?

Worksheet 9 (Continued)

7. "A distribution that is skewed to the right has a positive skew." Is this statement true or false?

8. A bimodal distribution has how many high points?

9. From left to right in a frequency polygon, are the scores arranged from high to low or from low to high?

Solution section:

two (it)   left (evil)   right (coffin)   three (pure)   false (be)

middle (there)   true (when)   high to low (flowers)   one (selfish)   frequencies (to)

scores (I)   positive (to)   normal (don't)   shape (everything)

lines (funeral)   negative (want)   low to high (happens)

Write the answer to the riddle here, putting one word on each line: _____ _____ _____ _____

_____ _____ _____ _____ _____

Notes:

Section 10: The Mean: An Average

The most popular average is the **mean**.[1] It is so popular that it is sometimes simply called *the average*; however, this is an ambiguous term because there are several different types of averages used in statistics.

Computation of the mean is quite simple; just sum (i.e., add up) the scores and divide by the number of scores. Here is an example:

Scores: 5, 6, 7, 10, 12, 15
Sum of scores: 55
Number of scores: 6
Computation of mean: 55/6 = 9.166 = 9.17

Notice in the example that the answer was computed to three decimal places and rounded to two. In scientific work, the mean is usually reported to two decimal places.

There are several symbols for the mean. In scientific journals, the most commonly used symbols are *M* and *m*.[2] Statisticians often use this symbol:

$$\bar{X}$$

It is pronounced *X-bar*. Remember that *X* without the bar stands for a score or a set of scores. Throughout this book, the symbol *M* will be used.

Although calculating the mean is simple, it is important to become familiar with the symbols in the formula because they will be used later in this book in other formulas. The formula for the mean is:

$$M = \frac{\Sigma X}{N}$$

---

[1] Its full, formal name is the *arithmetic mean*. Other averages are described in the next section.
[2] Strictly speaking, the upper-case *M* should be used when describing an entire population and the lower-case *m* should be used when describing a sample drawn from a population. Many authors of applied research, however, ignore this convention.

The symbol Σ, which is the Greek letter sigma, is pronounced *sum of* in statistics. The *X* stands for *score* or *scores*. *N* is the number of cases or subjects. Thus, the formula says: *The mean equals the sum of the scores divided by the number of cases.*

An important characteristic of the *mean* is that it is the balance point of the distribution; that is, it is the *point around which the deviations sum to zero.*[3]

The example in Table 10.1 illustrates this characteristic. The sum of the scores is 60; dividing this by 5 yields a mean of 12.00. By subtracting the mean from each score, we obtain the deviations from the mean. These deviations sum to zero.[4] (Notice the negatives cancel out the positives when summing.)

Table 10.1
Scores and Their Deviations from Their Mean

Score	Mean	Deviation
7	12.00	−5
11	12.00	−1
11	12.00	−1
14	12.00	2
17	12.00	5
	Sum of deviations = 0	

A major drawback of the mean is that it is drawn in the direction of extreme scores. This is a problem if there are *either* some extremely high scores that pull it up *or* some extremely low scores that pull it down. Here's an example expressed in cents of the contributions to charity by two groups of children:

Group A: 1, 1, 2, 3, 3, 4, 4, 4, 5, 5, 5, 5, 6, 6, 6, 7, 8, 10, 10, 10, 11
Mean for Group A = 5.52

---

[3]This is a *defining characteristic* of the mean. There is only one value that has this characteristic for a given distribution. Any value that does *not* have this characteristic is *not* the mean.
[4]If the mean is not a whole number, the sum of the deviations may vary slightly from zero because of rounding when determining the mean. A rounded mean is not *precisely* accurate.

Group B: 1, 2, 2, 3, 3, 3, 4, 4, 5, 5, 5, 6, 6, 6, 6, 6, 9, 10, 10, 150, 200
Mean for Group B = 21.24

Notice that, overall, the two distributions are quite similar. Yet the mean for Group B is much higher than the mean for Group A because of two students who gave extremely high contributions of 150 cents and 200 cents.[5] If only the means for the two groups were reported without reporting all of the individual contributions, it would suggest that the average student in Group B gave about 21 cents when, in fact, none of the students made a contribution of about this amount. An average that provides a more accurate indication of the center for this situation is described in the next section.

Another limitation of the mean is that it is only appropriate for use with *interval* and *ratio* scales of measurement (see Section 2) because its value is dependent upon the magnitude of the scores, which create the size of their deviations from the mean, unlike the averages described in the next section.

Note that a synonym for *average* is **measure of central tendency**. Although the latter term is seldom used in reports of scientific research, you may encounter it if you refer to other statistics texts.

Terms to Review Before Attempting Worksheet 10

mean, *M*, *m*, Σ, *X*, *N*, measure of central tendency

---

[5]A distribution with extremes that produce a tail in one direction but not the other is called *skewed*. See Section 9.

Worksheet 10: The Mean: An Average

Riddle: What type of magic can auto mechanics do?

DIRECTIONS: To find the answer to the riddle, write the answer to each question in the space immediately below it. The word in parentheses in the solution section next to the answer to the first question is the first word in the answer to the riddle, the word beside the answer to the second question is the second word, and so on.

1. To two decimal places, what is the mean of these scores: 10, 15, 17, 17, 20, 22?

2. To two decimal places, what is the mean of these scores: 0, 0, 2, 5, 8, 9, 12?

3. In scientific journals, what is the most common symbol for the mean?

4. In statistics, how is Σ pronounced?

5. For what does N stand?

6. If the mean for a group is 15.00 and Sylvia, who is a member of the group, obtained a score of 14, what is the value of the deviation score associated with Sylvia's score?

7. If the mean is subtracted from each of the scores underlying it and the deviations are summed, what value is obtained?

Worksheet 10 (Continued)

8. "For a skewed distribution, the mean is pulled in the direction of the extreme scores." Is this statement true or false?

9. Suppose that two homes at the top of a hillside community were sold for very high prices, and twenty homes lower on the hill were sold for modest prices. Would the mean probably be good for giving an accurate indication of the average home price in the community?

10. What is a synonym for the term *averages*?

Solution section:

X-bar (fix)   false (transmission)   true (underneath)   no (their)
measures of central tendency (car)   *X* (hat)   1 (hand)   16.83 (they)
zero (from)   number of cases (person's)   sum of (a)   sigma (seeing)
7.20 (lift)   5.14 (can)   14.43 (never)   *M* or *m* (pick)   −1 (pocket)

Write the answer to the riddle here, putting one word on each line: _____   _____   _____   _____ _____   _____   _____   _____   _____ _____

Notes:

Section 11: Mean, Median, and Mode

The *mean*, described in the previous section, is the *balance point* in a distribution. It is the most frequently used average.[1]

An alternative average is the *median*. It is the value in a distribution that has 50% of the cases above it and 50% of the cases below it. Thus, it is the *middle point* in a distribution. In Example 1, there are 11 scores. The middle score, with 50% on each side, is 81, which is the median. Note that there are five scores above 81 and five scores below 81.[2]

Example 1: Scores (arranged in order from low to high):
61, 61, 72, 77, 80, 81, 82, 85, 89, 90, 92

In Example 2, there are six scores. Because there is an even number of scores, the median is halfway between these scores. To find this value, sum the two middle scores (7 + 10 = 17) and divide by 2 (17/2 = 8.5). Thus, 8.5 is the median of this set of scores.

Example 2: Scores:
3, 3, 7, 10, 12, 15

An advantage of the median is that it is insensitive to extreme scores.[3] This is illustrated by Example 3, in which the extremely high score of 229 has no effect on the value of the median. The median is 8.5, the same value as in Example 2, despite the extremely high score.

Example 3: Scores:
3, 3, 7, 10, 12, 229

---

[1]Another term for *average* is *measure of central tendency*.

[2]When there are tie scores in the middle, that is, when the middle score is earned by more than one subject, this method for determining the median is only approximate and the result should be referred to as an *approximate median*.

[3]In Section 10, it was noted that the mean is pulled in the direction of extreme scores, which may make it a misleading average for skewed distributions.

The *mode* is another average. It is simply the *most frequently occurring score*. In Example 4, the mode is 7 because it occurs more often than any other score.

Example 4: Scores:
    2, 2, 4, 6, 7, 7, 7, 9, 10, 12

A disadvantage of the mode is that there may be more than one mode for a given distribution. This is the case in Example 5 in which both 20 and 23 are the mode.

Example 5: Scores:
    17, 19, 20, 20, 22, 23, 23, 28

Here are some guidelines to use when choosing an average:

➔ Other things being equal, choose the mean because more powerful statistical tests (described later in this book) can be applied to it than to the other averages. However: (1) The mean is appropriate only for approximately symmetrical distributions; it is inappropriate to use it to describe the average of a highly skewed distribution, and (2) the mean is only appropriate for describing interval and ratio data. (See Section 2 to review scales of measurement.)

➔ Choose the median when the mean is inappropriate. The exception to this guideline applies when describing nominal data. Nominal data (see Section 2) are naming data such as political affiliation, ethnicity, etc. Natural order is not inherent in nominal data; therefore, they cannot be put in an ordered sequence, which must be done in order to determine the median.

Remember that subjects must be ordered from low to high in order to calculate the median.

→ Choose the mode when an average is needed to describe nominal data. Note that when describing nominal data, it is often not necessary to use an average. For example, if there are more Democrats than Republicans in a community, the best way to describe this is to give the percentage of people registered in each party. To simply state that the modal political affiliation is Democratic (which is the average in this case) is much less informative than reporting percentages.

Note that in a perfectly symmetrical distribution (such as the normal distribution), the mean, median, and mode all have the same value. In skewed distributions, their values are different as illustrated in the following figures. In a distribution with a **positive skew**, the mean has the highest value because it is pulled in the direction of the extremely high scores. In a distribution with a **negative skew**, the mean has the lowest value because it is pulled in the direction of the extremely low scores. As noted earlier, do not use the mean when a distribution is highly skewed. See Table 11.1 for a review of these concepts.

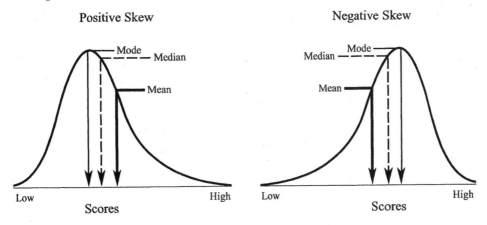

Figure 11.1. Positions of three averages in distributions with positive and negative skews.

Table 11.1
Comparison of the Three Averages

	Mean (M)	**Median (Mdn)**	**Mode (Mo)**
Definition	Balance point	Middle point	Most frequent score
When to Use	Symmetrical distributions of interval/ratio data	When mean is inappropriate (except for nominal data)	Nominal data
Frequency of Use in Scientific Reporting	Very frequent	Somewhat frequent	Very infrequent[1]
Relative Values for Positive Skew[2]	Higher than Mdn or Mo	Between M and Mo	Lower than M or Mdn
Relative Values for Negative Skew	Lower than Mdn or Mo	Between M and Mo	Higher than M or Mdn

[1]Although nominal data (for which the mode is appropriate) is frequently reported in scientific writing, most researchers report percentages for each category rather than reporting the modal category. Thus, the mode is seldom used.

[2]See Section 9 to review skewed distributions.

Terms to Review Before Attempting Worksheet 11

mean, median, mode, positive skew, negative skew

Worksheet 11: Mean, Median, and Mode

> ### *Riddle*: How are political speeches similar to food?

DIRECTIONS: To find the answer to the riddle, write the answer to each question in the space immediately below it. The word in parentheses in the solution section next to the answer to the first question is the first word in the answer to the riddle, the word beside the answer to the second question is the second word, and so on.

1. Which average is defined as the value that has 50% of the cases below it?

2. What is the median of these scores: 0, 5, 7, 9, 9?

3. What is the median of these scores: 10, 12, 14, 17?

4. What is the median of these scores: 50, 40, 29, 52, 54?

5. What is the mode of these scores: 11, 12, 12, 15, 18, 20?

6. Which average is appropriate only for describing interval and ratio data?

7. Should the mean be used to describe highly skewed distributions?

8. Which average is most appropriate for describing nominal data?

Worksheet 11 (Continued)

9. Do the mean, median, and mode all have the same value in a normal distribution?

10. "In a distribution with a positive skew, the mean has a higher value than the median." Is this statement true or false?

11. "In a distribution with a negative skew, the median has a lower value than the mean." Is this statement true or false?

Solution section:

8 (win)	9 (election)	12.5 (hungry)	median (very)	13 (they)
7 (often)	mean (baloney)	false (thought)	39 (being)	12 (just) 50 (are)
true (for)	yes (food)	no (disguised)	51 (kitchen)	14 (weakness)
7.5 (milk)	averages (windfall)	9.5 (sensitivity)	mode (as)	

Write the answer to the riddle here, putting one word on each line: _____ _____ _____ _____

_____ _____ _____ _____ _____

_____ _____

Section 12: Variability: The Range and Interquartile Range

Variability refers to the differences among scores, which indicate how subjects vary.[1] In scientific writing, most authors report a statistic designed to indicate the amount of variability[2] immediately after reporting an average.

How much subjects in a group vary is important for statistical and practical reasons. Suppose, for example, you were going to teach fourth grade next year and were offered a choice between two classes—both of which were very similar in terms of their *average* scores obtained on standardized tests. Before making a choice, you would be wise to ask about the variability. Suppose you learned that one class had very little variability—their percentile ranks were all very close to their average—and that the other had tremendous variability—their percentile ranks *varied* from the highest to the lowest score possible with a great deal of spread in between. Which class would you choose? Clearly, information on variability would be important in helping you make a decision.

A simple statistic that describes variability is the *range*. It is computed by subtracting the lowest score from the highest score.[3] For the scores in Example 1, the range is 18 (20 minus 2). We could report 18 as the range or simply state that the scores range from 2 to 20.

Example 1: Scores:
> 2, 5, 7, 7, 8, 8, 10, 12, 12, 15, 17, 20

A weakness of the range is that it is based on only two scores, which may not reflect the true variability of all the scores. Consider Example 2 in which the range is also 18. Note, however, that there is actually very little variability among the subjects; one subject with an extreme score has had an undue influence on the range.

---

[1] Synonyms for *variability* are *spread* and *dispersion*.
[2] This group of statistics is often referred to as *measures of variability*.
[3] Some statisticians add the constant 1 to the difference when computing the range.

Example 2: Scores:

2, 2, 2, 3, 4, 4, 5, 5, 5, 6, 7, 20

Scores such as the 20 in Example 2 are known as ***outliers***. They lie far outside the range of the vast majority of the scores.

A better measure of variability is the ***interquartile range*** (***IQR***). It is defined as the range of the middle 50% of the subjects. To find the interquartile range, do the following:

1. Put the scores in order from low to high. Then determine how many scores constitute *one-quarter* of the scores. In Example 3, there are 12 scores; one-quarter of them (12 divided by 4) equals 3.

2. Count up from the lowest score the number of scores you calculated in Step 1. In Example 3, when you count up 3 scores you come to the arrow on the left, which is between the scores of 2 and 3. Halfway between them is a value of 2.5.

3. Count down from the highest score the number of scores you calculated in Step 1. In Example 3, when you count down three scores, you come to the arrow on the right, which is between the scores of 5 and 6. Halfway between them is a value of 5.5.

4. Subtract the answer to Step 2 from the answer to Step 3 (5.5 − 2.5 = 3.0). Thus, 3.0 is the value of the *interquartile range* for this set of scores.[4] When you report 3.0 to an audience, they will know that the range of the middle 50% of the subjects is only 3 points. Note that the undue influence of the outlier of 20 has been overcome by using the *interquartile range* instead of the *range*.

Example 3: Scores:

2, 2, 2, 3, 4, 4, 5, 5, 5, 6, 7, 20
⇑   ⇑

---

[4] This procedure is approximate when there are tie scores at the points at which you are working. When this is the case, the answer should be reported as the *approximate interquartile range*.

The interquartile range may be thought of as a first cousin of the *median*. (Remember that to calculate the median, you count to the middle of the distribution.) Thus, when the *median* is reported as the average for a set of scores, it is customary to report the *interquartile range* as the measure of variability.[5]

Terms to Review Before Attempting Worksheet 12

variability, range, outliers, interquartile range

"You've been working awfully hard on your statistics homework. If you need a little fresh air and sunshine, you can go to www.fresh-air-and-sunshine.com"

---

[5]The measure of variability that is associated with the mean is introduced in the next section. See the previous section for guidelines on when to report the median and the mean.

Worksheet 12: Variability: The Range and Interquartile Range

Riddle: According to Sam Ewing, who are your best friends?

DIRECTIONS: To find the answer to the riddle, write the answer to each question in the space immediately below it. The word in parentheses in the solution section next to the answer to the first question is the first word in the answer to the riddle, the word beside the answer to the second question is the second word, and so on.

1. *Spread* and *dispersion* are synonyms for what term?

2. If two groups are equal on the average, will they necessarily be equal in their variability?

3. What is the range of these scores: 6, 8, 8, 10, 12, 13?

4. What is the outlier in this set of scores: 2, 10, 10, 11, 13, 14, 15?

5. What percentage of the subjects is encompassed by the interquartile range?

6. Is the interquartile range or the range a better measure of variability?

Worksheet 12 (Continued)

7. What is the interquartile range of these scores: 10, 10, 11, 12, 13, 13, 14, 14, 15, 17, 18, 18?

8. What is the interquartile range of these scores: 5, 5, 0, 1, 2, 11, 11, 11, 10, 9, 7, 6, 5, 5, 0, 4?

9. With which average is the interquartile range associated?

Solution section:

median (back)   6.5 (your)   15 (young)   6 (smile)   mean (fell)

interquartile range (you)   4.5 (behind)   5 (friendly)

mode (being)   68% (figure)   yes (loving)   variability (those)

2 (well)   7 (speak)   no (who)   50% (of)   99% (brilliant)   95% (calling)

Write the answer to the riddle here, putting one word on each line: _____ _____ _____ _____ _____ _____ _____ _____ _____

Notes:

Section 13: Variability: Introduction to the Standard Deviation

The ***standard deviation*** is a measure of how much scores differ (or *vary*) from their mean—the average that was introduced in Section 10. By reporting just the mean and standard deviation of a set of scores, an author is usually able to convey a good picture of a distribution.

Statisticians use the symbol S for the standard deviation of the scores of a population.[1] Authors of applied research in journals sometimes use the symbol ***S.D.***—sometimes with and sometimes without the period marks.

The formula that defines the standard deviation is:

$$S = \sqrt{\frac{\Sigma x^2}{N}}$$

The lower-case x stands for the deviation of a score from the mean of its distribution. To obtain the deviations for Example 1, first calculate the mean (in this case, $78/6 = 13.00$) and subtract the mean from each score, as shown below. Then square the deviations and sum the squares, as indicated by the symbol Σ. Then enter this value in the formula along with the number of cases (N) and perform the calculations as illustrated at the top of the next page.

Example 1:

Scores (X)	Deviations ($X - M$)	Deviations Squared x^2
10	$10 - 13.00 = -3$	9.00
11	$11 - 13.00 = -2$	4.00
11	$11 - 13.00 = -2$	4.00
13	$13 - 13.00 = 0$	0.00
14	$14 - 13.00 = 1$	1.00
19	$19 - 13.00 = 6$	36.00
		$\Sigma x^2 = 54.00$

---

[1]The formula for the standard deviation of a population is given in this section. The formula for estimating the standard deviation of a population from a sample is given in Appendix A.

Thus,

$$S = \sqrt{\frac{54.00}{6}} = \sqrt{9.00} = 3.00$$

An algebraically equivalent formula, known as the computational formula, which is slightly easier to use, is presented in Appendix A.[2] It is advantageous to use this formula when you have a large number of scores and are using a calculator. Of course, if you have a very large number of scores, a computer is recommended for calculations.

Considering how the standard deviation is calculated should give you a feeling for the meaning of the standard deviation. As the formula indicates, it is the *square root of the average squared deviation from the mean.* Thus, the larger the deviations from the mean, the larger the standard deviation. Conversely, the smaller the deviations from the mean, the smaller the standard deviation. At the extreme, when all the scores are the same, the standard deviation equals zero. For example, if each subject in a population had a score of 20, then the mean of the scores would be 20.00. When the mean of 20.00 is subtracted from each score of 20, the resulting deviations will all be zero; when they are squared, they will equal zero; when they are summed, they will equal zero; when zero is entered into the numerator of the formula, the solution for *S* will be zero. This is as it should be; when there is no variation, *S* equals zero—indicating there is no variation.

The standard deviation takes on a special meaning when considered in relation to the normal curve (see Section 9) because it was designed expressly to describe this distribution. Here is a simple rule to remember: *About two-thirds of the cases lie within one standard deviation unit of the mean in a normal distribution.* (Note that "within one standard deviation unit" means one unit on *both* sides of the mean.) For example, suppose that the mean of a set of normally distributed scores equals 70.00 and the standard deviation equals 10.00.

---

[2]The *definition* formula is presented in this section because it is the one that most readily conveys an understanding of what influences the size of standard deviation for a set of scores.

Then, about two-thirds of the cases lie within 10 points of the mean. More precisely, 68% of the cases lie within 10 points of the mean as illustrated in Figure 13.1.

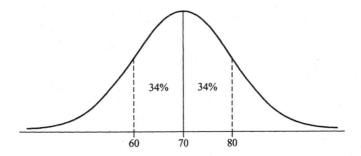

Figure 13.1. Normal curve with a standard deviation of 10.00.

Suppose for another group, the mean of their normal distribution also equals 70.00, but their standard deviation equals 5.00. Then, 68% of the cases lie within 5 points of the mean as illustrated in Figure 13.2.

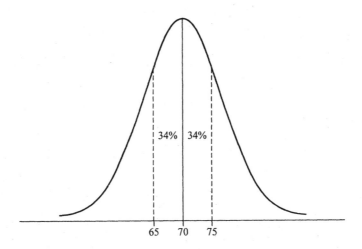

Figure 13.2. Normal curve with a standard deviation of 5.00.

At first, this seems like magic—regardless of the value of the standard deviation, 68% of the cases lie within one standard deviation unit in a normal curve. Actually, it is not magic but a property of the normal curve. When you are calculating the standard deviation, you are actually calculating the number of points that one must go out from the mean to capture the middle 68% of the cases.[3] This two-thirds rule does *not* strictly apply if the distribution is *not* normal. The less normal it is, the less accurate the rule is. Other rules are described in the next section.

Don't forget that the standard deviation is a first cousin of the mean. Thus, when researchers report the mean (the most popular average), they usually also report the standard deviation.

Terms to Review Before Attempting Worksheet 13

standard deviation, *S*, *S.D.*

---

[3]When you go out one standard deviation on both sides of the mean, you reach the *points of inflection*—the points where the curve changes direction and begins to go out more quickly than it goes down.

Worksheet 13: Variability: Introduction to the Standard Deviation

---

Riddle: According to Abraham Lincoln, how are vices and virtues related?

DIRECTIONS: To find the answer to the riddle, write the answer to each question in the space immediately below it. The word in parentheses in the solution section next to the answer to the first question is the first word in the answer to the riddle, the word beside the answer to the second question is the second word, and so on.

1. Statisticians use what symbol for the standard deviation?

2. What is the symbol for the deviation of a score from its mean?

3. To two decimal places, what is the standard deviation of these scores: 3, 5, 6, 7, 8?

4. To two decimal places, what is the standard deviation of these scores: 0, 2, 5, 9, 12?

5. According to the standard deviations, are the scores in "Question 3" *or* "Question 4" more variable?

6. If everyone in a population has the same score, what is the value of the standard deviation for their scores?

Worksheet 13 (Continued)

7. The standard deviation was designed expressly to describe what type of distribution?

8. If the mean for a group equals 40.00 and the standard deviation equals 5.00, what percentage of the cases in a normal distribution lies between 35 and 45?

9. If the mean for a group equals 27.00 and the standard deviation equals 3.00, what percentage of the cases in a normal distribution lies between 24 and 27?

Solution section:

68% (few) X (evil)   34% (virtues)   2.93 (weary)   4.70 (slavery)

99% (shift)   zero (have)   normal (very) S (folks)   1.72 (have)

x (who)   4.41 (no)   scores in question 3 (sleepless)   50% (helpful)

scores in question 4 (vices)   skewed (being) Σ (goodness)

Write the answer to the riddle here, putting one word on each line: _____ _____ _____ _____
_____ _____ _____ _____ _____

Section 14: A Closer Look at the Standard Deviation

In the previous section, you learned that the *standard deviation* is a measure of how much scores vary from their mean. More specifically, you learned that about 68% of the scores in a normal distribution lie within one standard deviation unit of the mean. In this section, you'll learn some additional rules for interpreting the standard deviation.[1]

The *approximate 95% rule* says that if you go out 2 standard deviation units on both sides of the mean in a normal distribution, you will find approximately 95% of the cases. Here's an example of the application of the approximate 95% rule:

Example 1:
The mean for a group equals 35.00 and the standard deviation equals 6.00. Two standard deviation units equal 12.00 points (2 × 6.00 = 12.00). Thus, if you (a) go up 12 points from the mean (35.00 + 12.00 = 47.00) and (b) go down 12 points from the mean (35.00 – 12.00 = 23.00), you have identified the scores (47.00 and 23.00) between which approximately 95% of the cases lie.

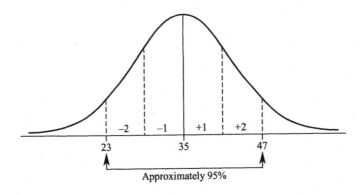

Figure 14.1. Normal curve illustrating 95% rule.

---

[1]The rules are derived from the table of the normal curve, which is introduced in a later section.

The **_99.7% rule_** says that if we go up and down 3 standard deviation units from the mean, we will find 99.7% of the cases. For the information in Example 1, multiply 3 times the standard deviation (3 × 6.00 = 18.00). Going up and down 18 points from the mean yields these scores: 53.00 and 17.00. This is illustrated here:

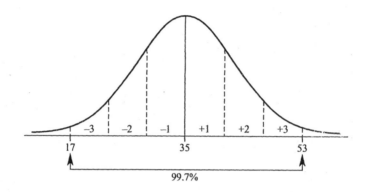

Figure 14.2. Normal curve illustrating 99.7% rule.

Let's review:

If *M* = 35.00 and *S* = 6.00, then:
      (1) 68% of the cases lie between 29.00 and 41.00;
      (2) 95% of the cases lie between 23.00 and 47.00;
      (3) 99.7% of the cases lie between 17.00 and 53.00.

You can see that almost all the cases (99.7%) in a normal distribution lie within 3 standard deviation units of the mean. Thus, for practical purposes we can say that a normal distribution has only 6 standard deviation units—3 above the mean and 3 below the mean.

Don't lose sight of the fact that we are examining the standard deviation to determine the variability in a set of scores. To illustrate this, consider Example 2. The mean is the same as in Example 1, but the standard deviation is only half its size. Results of applying the three rules are shown. By comparing the two examples, it becomes obvious that the smaller the standard deviation, the less far out you need to go to capture a given percentage of cases. In fact, with a standard deviation of only 3.00, you need to go only to 26.00 and 44.00 to capture 99.7% of the cases.

Example 2:

When $M = 35.00$ and $S = 3.00$, then approximately:

(1) 68% of the cases lie between 32.00 and 38.00;

(2) 95% of the cases lie between 29.00 and 41.00;

(3) 99.7% of the cases lie between 26.00 and 44.00.

If you were reporting on research in which the groups in Example 1 and Example 2 were being compared, you could say that the two groups are equal on the average (as measured by the mean) but that the first group has twice the variability (as measured by the standard deviation) than the second group. Potentially, this could be very important information for your readers.

When doing the following worksheet, keep the following multipliers in mind. Multiply the standard deviation for a set of scores by the appropriate multiplier before adding and subtracting from the mean.

Rule	Multiplier
68%	1
95%	2
99.7%	3

Additional rules are discussed in the next section. Remember, all of these rules strictly apply only in the case of a normal distribution.

Terms to Review Before Attempting Worksheet 14

approximate 95% rule, 99.7% rule

Worksheet 14: A Closer Look at the Standard Deviation

> *Riddle*: When do most people stop believing in heredity?

DIRECTIONS: To find the answer to the riddle, write the answer to each question in the space immediately below it. The word in parentheses in the solution section next to the answer to the first question is the first word in the answer to the riddle, the word beside the answer to the second question is the second word, and so on.

1. If you go out 1 standard deviation unit on both sides of the mean in a normal distribution, what percentage of the cases will you capture?

2. If you go out 3 standard deviation units on both sides of the mean in a normal distribution, what percentage of the cases will you capture?

3. For $M = 100.00$ and $S = 10.00$, approximately what percentage of the cases lie between 80.00 and 120.00?

4. For $M = 55.00$ and $S = 7.00$, between what two values do approximately 95% of the cases lie in a normal distribution?

5. For $M = 90.00$ and $S = 15.00$, between what two values do 99.7% of the cases lie in a normal distribution?

Worksheet 14 (Continued)

6. What is the multiplier for the approximate 95% rule?

7. What is the multiplier for the 99.7% rule?

8. For all practical purposes, the normal curve has how many standard deviation units?

Solution section:

1 (birth)	4 (grandparents)	51.30–128.70 (years)	2 (act)
6 (delinquents)	68% (when)	99% (becoming)	99.7% (their)
95% (children)	48.00–62.00 (useful)	3 (like)	41.00–69.00 (start)
50% (happening)	14.00–28.00 (wild)	34% (heaven)	45.00–135.00 (to)

Write the answer to the riddle here, putting one word on each line: _____ _____ _____ _____
_____ _____ _____ _____

Notes:

Section 15: Another Look at the Standard Deviation

When the standard deviation is reported in conjunction with the mean of a set of scores, it gives an indication of the amount of variability in the distribution. For example, suppose you read the following statement in a report:

> For the population of Martians, the mean score on the
> Human Awareness Scale is 44.00 and the standard de-
> viation is 4.00. The distribution is normal.

From this, you should be able to picture in your mind's eye that about 68% of the Martians had scores between 40.00 and 48.00.[1] Put another way, the vast majority of the Martians had scores within 4 points of the mean. This indicates the amount of variability because it indicates the number of points within which the vast majority fall.

In Section 14, you learned that *approximately* 95% of the cases lie within 2 standard deviation units of the mean. To be more precise, the ***precise 95% rule*** says that if you go out 1.96 standard deviation units from the mean in a normal distribution, you will find 95% of the cases.[2] The constant 1.96 is derived from the definition of the normal curve.[3] To some students, at first this seems a little like magic. It might help to keep this in mind: If 95% of the cases do *not* lie within 1.96 standard deviation units of the mean, the distribution is *not* normal. Thus, this rule applies to all normal distributions, regardless of the value of their means and standard deviations.

Let's apply the precise 95% rule to the Martian example:
1. Multiply 1.96 times S ($1.96 \times 4.00 = 7.84$).
2. Subtract the result of Step 1 from the mean ($44.00 - 7.84 = 36.16$).
3. Add the result of Step 1 to the mean ($44.00 + 7.84 = 51.84$).

---

[1]To review this rule, see Section 13.
[2]Note that 1.96 is very close to 2, which led to the approximate rule.
[3]The table of the normal curve is introduced in Section 18. When we examine it, you will again encounter the constant 1.96.

The results of Steps 2 and 3 yield the points between which 95% of the cases lie; thus, 95% of the cases in this normal distribution lie between 36.16 and 51.84.

The **99% rule** says that if you go out 2.58 standard deviation units from the mean, you will identify the values between which 99% of the cases lie.[4] Let's apply the 99% rule to the Martian example:

1. Multiply 2.58 times S ($2.58 \times 4.00 = 10.32$).
2. Subtract the result of Step 1 from the mean ($44.00 - 10.32 = 33.68$).
3. Add the result of Step 1 to the mean ($44.00 + 10.32 = 54.32$).

The results of Steps 2 and 3 yield the points between which 99% of the cases lie; thus, 99% of the cases in this normal distribution lie between 33.68 and 54.32.

The standard deviation has been examined in detail not only because it is useful in describing the variability in a group but also because it is used in a variety of other statistical procedures. These procedures will be much easier to master if you feel thoroughly comfortable with the standard deviation.

Terms to Review Before Attempting Worksheet 15

precise 95% rule, 99% rule

---

[4] In the previous section, you learned that if you go out 3 standard deviation units from the mean, you identify the values between which 99.7% of the cases lie. As you will see later in this book, 99% is of more interest than 99.7% for advanced statistical work.

Worksheet 15: Another Look at the Standard Deviation

> *Riddle*: What happens if a wealthy person dies without a will?

DIRECTIONS: To find the answer to the riddle, write the answer to each question in the space immediately below it. The word in parentheses in the solution section next to the answer to the first question is the first word in the answer to the riddle, the word beside the answer to the second question is the second word, and so on.

1. If $M = 500.00$ and $S = 100.00$ in a normal distribution, what percentage of the subjects have scores between 400.00 and 600.00?

2. According to the precise 95% rule, how many standard deviation units should you go out from the mean in order to capture 95% of a population in a normal distribution?

3. Does the precise 95% rule apply to all normal distributions?

4. In a normal distribution with $M = 78.56$ and $S = 12.92$, between what two values does the middle 95% of a population lie according to the precise rule?

5. In a normal distribution with $M = 50.00$ and $S = 8.00$, what percentage of the population lies between 29.36 and 70.64?

Worksheet 15 (Continued)

6. In a normal distribution with $M = 150.00$ and $S = 20.00$, what percentage of the population lies between 110.80 and 189.20?

7. In a normal distribution with $M = 11.52$ and $S = 1.40$, between what two values does 99% of the population lie?

8. If Group A has $M = 52.39$ ($S = 3.44$) and Group B has $M = 41.55$ ($S = 4.19$), which group has greater variability?

Solution section:

Group B (heirs)   7.91–15.13 (his or her)   no (visible)   2.58 (death)

99% (sides)   95% (become)   Group A (estate)   10.12–12.92 (feeling)

68% (the)   yes (on)   99.7% (funeral)   34% (jumping)   65.64–91.48 (wealth)

1.96 (lawyers)   34.00–66.00 (wishing)   53.24–103.88 (both)   average (dispute)

Write the answer to the riddle here, putting one word on each line: _____ _____ _____ _____
_____ _____ _____ _____

Section 16: Standard Scores

Up to this point, the emphasis has been on describing a distribution in order to get an overview of the data for a group as a whole—either by using a table such as a frequency distribution, a figure such as a histogram, or by reporting an average and a measure of variability. In this section and the next, some methods for describing where an *individual* stands in a group will be considered.

Raw scores (i.e., the number of points earned) are often reported back to individuals. Raw scores are obtained from many kinds of measures—from multiple choice tests to instruments that measure blood pressure—from a point system for creativity in an artistic endeavor to essay examinations. Sometimes, raw scores are easily interpreted because we have a frame of reference for them. For example, a person who has been monitoring her blood pressure over time probably already knows what the normal range is and what values might be dangerous. Such a person is engaging in a *norm-referenced interpretation.* That is, the norms for normal, below average, and above average, based on what has been commonly observed among large numbers of patients, provide reference points for interpreting raw scores. Frequently, however, raw scores have little meaning to the recipient. Suppose, for example, you took a new paper-and-pencil depression scale and were told that your raw score was 56. What does this mean? Very little without additional information.

Standard scores (frequently called *z*-scores) help in interpreting raw scores by indicating where an individual stands in a group. Specifically, a standard score indicates *how many standard deviation units a person is from the mean and whether the person is above or below the mean.* For example, suppose a person was told that she had a *z*-score of 1.00. This means that she is exactly one standard deviation unit from the mean. Furthermore, because a positive *z*-score indicates that a person is above the mean and a negative *z*-score indicates that a person is below the mean, we know that she is above the mean. Drawing on our knowledge of the normal curve, we also know that a *z*-score of 1.00 puts her above about 84% of the subjects in a normal distribution. This is how we know this: First, 50% of the subjects lie below the mean in a normal distribution;[1] and second, about 34% of the cases lie between the mean and one standard deviation unit above the mean. This is illustrated here:

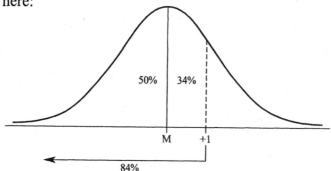

Figure 16.1. Normal curve illustrating percentage below a *z*-score of 1.00.

Suppose that another person had a *z*-score of 0.00. This indicates that the person has a score whose value is the same as the value of the mean (i.e., zero standard deviations above or below the mean).

Finally, suppose that someone had a *z*-score of –2.00. This indicates that the person is 2 standard deviation units below the mean.

Because 3 standard deviation units on both sides of the mean encompass 99.7% of the cases in a normal curve, *z*-scores seldom are higher than 3.00 or lower than –3.00. This illustrates the effective range of *z*-scores:

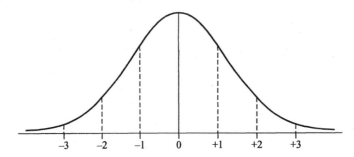

Figure 16.2. Normal curve illustrating the effective range of *z*-scores.

---

[1]Remember that the normal distribution is perfectly symmetrical and the mean is in the middle. Therefore, 50% of the cases are above the mean and 50% are below the mean.

Of course, people often obtain intermediate values between those shown in Figure 16.2.

To determine an individual's z-score, apply this formula:

$$z = \frac{X - M}{S}$$

Before using the formula, consider what it means. The numerator says to subtract the mean from a person's score; this is how many raw score points a person's score is from the mean. By dividing this difference by the standard deviation, we are determining *how many standard deviation units* a person is from the mean.

Suppose that a child weighs 55.00 pounds and the mean for her age, height, and gender is 60.00 pounds and the standard deviation is 5.00. Entering this information in the formula, we find that she is one standard deviation unit below the mean, which is much more meaningful than just knowing her weight expressed in pounds. Note that pounds are raw scores.

$$z = \frac{55.00 - 60.00}{5.00} = \frac{-5.00}{5.00} = -1.00$$

Consider this example:

Example 1:

$M = 75.00$, $S = 5.00$ for a group on the Weirdness Scale.

Here are some of the scores:

Subject	Raw Score	z-score
Jean	80	1.00
Tom	75	0.00
Sam	65	–2.00
Alice	62	–2.60

and other subjects not shown here.

In Example 1, notice that Jean is 5 raw score points above the mean; since 5 is the value of the standard deviation, she is one standard deviation above the mean, giving her a z-score of 1.00. Tom is right at the mean; therefore, he has a z-score of 0.00. Sam is 10 raw score points below the mean; because this is 2 standard deviation units below the mean ($2 \times 5.00 = 10.00$), he has a z-score of -2.00. The z-score for Alice illustrates that not everyone has a whole number as a z-score; because of this, z-scores should be computed to one or two decimal places.

It is quite possible that you cannot recall ever being given a z-score for your performance on a task. If they are so useful, why have they not been reported to you? Actually, they probably have been, but only after being transformed to another scale before being reported. The transformation of z-scores into other scales is taken up in the next section.

Terms to Review Before Attempting Worksheet 16

raw scores, standard scores, z-scores

"I'm the consultant they brought in to create some new statistical buzzwords."

Worksheet 16: Standard Scores

> *Riddle*: What will the ultimate economy car of the future do?

DIRECTIONS: To find the answer to the riddle, write the answer to each question in the space immediately below it. The word in parentheses in the solution section next to the answer to the first question is the first word in the answer to the riddle, the word beside the answer to the second question is the second word, and so on.

1. What term is used to refer to the number of points earned by an individual?

2. What is another term for *z*-scores?

3. If a person has a *z*-score of 1.50, how many standard deviation units is that person from the mean of the group?

4. A person with a *z*-score of 1.00 has a score that is higher than what percentage of the group in a normal distribution?

5. A person with a *z*-score of 0.00 has a score that is higher than what percentage of the group in a normal distribution?

6. According to this topic, *z*-scores are seldom lower than what value?

Worksheet 16 (Continued)

7. What is the highest value in the effective range of z-scores?

8. If $M = 42.35$ and $S = 2.87$ and June has a raw score of 45, what is the value of her z-score?

9. If $M = 95.00$ and $S = 5.00$ and Jake has a raw score of 90, what is the value of his z-score?

10. If $M = 511.12$ and $S = 20.28$ and Sarah has a raw score of 460, what is the value of her z-score?

11. Examine questions 8, 9, and 10. Notice that Sarah has the highest raw score. Who has the highest z-score?

Solution section:

490.84 (less)   39.48 (jump)   90.00 (dealer)   Sarah (being)   1.50 (just)

84% (sit)   raw score (it)   Jake (cost) z-score (gasoline)   15 (useful)

64% (start)   June (neighbors)   −3.00 (the)   0.92 (and)   34% (oil)

standard scores (will)   50% (in)   0.00 (garage)   −1.00 (impress)

3.00 (driveway)   −2.52 (the)   scores (speedy)   99.7% (Detroit)   95% (list)

Worksheet 16 (Continued)

Write the answer to the riddle here, putting one word on each line: _____ _____ _____ _____

_____ _____ _____ _____ _____

_____ _____

"He's determined to be above the
mean, the median, and the mode!"

Notes:

Section 17: Transformed Standard Scores

Standard scores (i.e., *z*-scores) indicate how many standard deviations a subject is from the mean of his or her group. At the bottom of page 86, you can see that, for all practical purposes, they range from –3.00 to 3.00. A *z*-score of 0.00 indicates that a subject is at the mean—that is, exactly average. Although these scores are quite understandable to those who have studied statistics, they are likely to be confusing to a layperson. Consider a person who knows she is about average and has worked hard on a test; when told that her score is 0.00, she is likely to be confused. Another person who is about average and obtains a *z*-score of –.10 (a tenth of a standard deviation below the mean) would probably be confused by his negative score.

To get around the undesirable features of *z*-scores, they are usually transformed to another scale that does not have negatives. These scores will be referred to here as ***transformed standard scores*** (***TSS***).[1]

A statistician named McCall suggested the following transformation, which he called a ***T score***:

$$T = (z)(10) + 50$$

As you can see, the transformation is quite simple. To calculate a set of *T* scores, simply multiply each person's *z*-score by 10 and add 50. Let's see what happens to the score of a person who has a *z*-score of 0.00 when we apply the transformation:[2]

$$T = (0.00)(10) + 50 = 0.00 + 50 = 50.00 = 50$$

Notice that the constant that is added (50) becomes the *new mean* of the set of scores. Keep in mind that the person who had a *z*-score of 0.00 was exactly at the mean; this person now has a *T* score of 50; the person is still exactly at the

---

[1]There is no standard symbol for transformed standard scores. For the sake of brevity, the acronym *TSS* is used here.
[2]Note that *TSS* are usually reported as whole numbers.

mean after all of the scores have been transformed. Because all of the scores in a set are transformed, subjects do not change positions in the group; only their scores change to a new scale.

Let's transform a z-score of –1.00 to a T score:

$$T = (-1.00)(10) + 50 = -10.00 + 50 = 40.00 = 40$$

Remember that a z-score of –1.00 indicates that the person is one standard deviation unit below the mean; the person remains at this position but now has a T score of 40. Thus, 40 is one standard deviation unit below the mean.

Since 50 is the mean of a set of T scores and 40 is one standard deviation unit below the mean, 10 (the difference between 50 and 40) must be the value of the *new standard deviation*. In Figure 17.1, this is illustrated with a normal curve, although the distribution does *not* have to be normal for the transformation to apply.

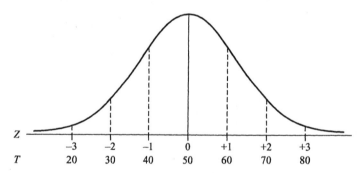

Figure 17.1. Normal curve with z-scores and equivalent T-scores.

A variety of other transformations have been developed. Here is a common one. To obtain a set of *IQ scores* use this formula:[3]

$$IQ = (z)(15) + 100$$

You should be able to see that a person with a z-score of 0.00 will have an *IQ* of 100 and a person with a z-score of 1.00 will have an *IQ* of 115. Thus, the

---

[3]On some *IQ* tests, a multiplier of 16 is used.

mean of a set of *IQ* scores for the norm group is transformed to 100 and the standard deviation is transformed to 15, as illustrated in Figure 17.2.

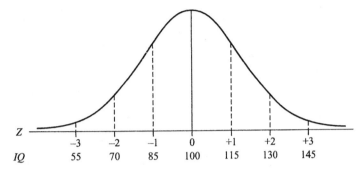

Figure 17.2. Normal curve with *z*-scores and equivalent *IQ* scores.

Another widely known transformation is the one originally used by the College Entrance Examination Board (CEEB) for their *Scholastic Aptitude Test (SAT)*. It was applied separately for the verbal and quantitative tests as shown here, which made the mean equal to 500 and the standard deviation equal to 100. The **CEEB scores** were derived as follows:

$$CEEB = (z)(100) + 500$$

For a person with a *z*-score of 1.50, her *CEEB* score was derived as follows:

$$CEEB = (1.50)(100) + 500 = 150 + 500 = 650$$

Figure 17.3 illustrates the effective range of *CEEB* scores:

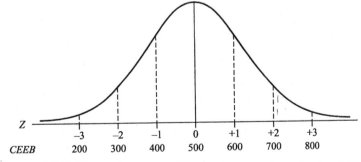

Figure 17.3. Normal curve with *z*-scores and equivalent *CEEB* scores.

A test-maker may, of course, select any two constants to use in the formula for *transformed standard scores*. The generalized formula for obtaining them is:

$$TSS = (z)(\text{new standard deviation}) + (\text{new mean})$$

Keep in mind that whatever constant is multiplied by all of the *z*-scores will become the new standard deviation, and whatever constant is added to the product will become the new mean.

Terms to Review Before Attempting Worksheet 17

transformed standard scores, *T* score, *IQ* scores, *CEEB* scores

**"Statistics class will be important to you
later in life because there's going to
be a test six weeks from now."**

Worksheet 17: Transformed Standard Scores

Riddle: According to Napoleon, what is the problem with the credit assigned for winning battles?

DIRECTIONS: To find the answer to the riddle, write the answer to each question in the space immediately below it. The word in parentheses in the solution section next to the answer to the first question is the first word in the answer to the riddle, the word beside the answer to the second question is the second word, and so on.

1. What is the mean of a set of T scores?

2. If $M = 35.00$ and $S = 5.00$ for a set of raw scores, what is Stan's T score if he has a raw score of 30?

3. In Question 2, how many standard deviation units is Stan from the mean?

4. If $M = 95.32$ and $S = 8.88$ for a set of raw scores, what is June's T score if she has a raw score of 112?

5. What is the mean of a set of IQ scores?

6. If $M = 49.55$ and $S = 5.11$ for a set of raw scores, what is Lyle's IQ score if he has a raw score of 45?

Worksheet 17 (Continued)

7. If $M = 66.47$ and $S = 9.68$ for a set of raw scores, what is Jennifer's *CEEB* score if she has a raw score of 85?

8. In terms of standard deviation units, which of these is the highest score: $T = 70$, $IQ = 105$, $CEEB = 466$?

9. If $M = 22.00$ and $S = 4.00$ and if Fernando has a z-score of 1.00, what is his *TSS* on a scale with a new mean of 300 and a new standard deviation of 50?

Solution section:

40 (win)   691 (get)   10 (keeping) $IQ = 105$ (war)   350 (credit)
69 (but) $CEEB = 466$ (famous)   113 (wishing) $T = 70$ (the)
5 (bloody)   50 (soldiers)   1 (battles)   87 (generals)   100 (the)   115 (fight)

Write the answer to the riddle here, putting one word on each line: _____ _____ _____ _____

_____ _____ _____ _____ _____

Section 18: Standard Scores and the Normal Curve

For any given standard score (z-score), we can determine the percentage of cases between it and the mean of a normal curve. We can also determine the percentage of cases above and below the z-score. You probably remember some of them. For example, for a z-score of 1.00, about 34% of the cases are between the score and the mean—more precisely, 34.13%. You can confirm this by looking up a z-score of 1.00 in the **table of the normal curve** in Table 1 near the end of this book. In the second column, you will find 34.13. In the third column, you will find 84.13, which is the percentage of cases below 1.00. In the fourth column, you will find 15.87, which is the percentage of cases above 1.00. Figure 18.1 illustrates this information:

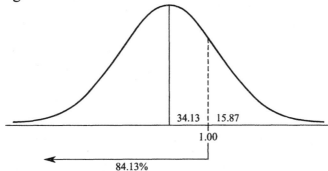

Figure 18.1. Percentage of cases below a z-score of 1.00.

For a z-score of –1.00, you use the same information from the table, keeping in mind that you are considering a negative standard score. This is illustrated in Figure 18.2.

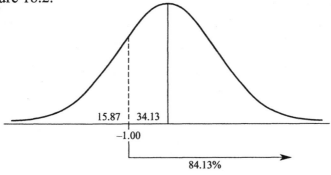

Figure 18.2. Percentage of cases above a z-score of –1.00.

You may recall the multiplier of 1.96, which we used when interpreting the standard deviation. Remember that if you go out 1.96 standard deviation units in both directions from the mean, you capture 95% of the cases. We can confirm this by looking up a z-score of 1.96 in Table 1.[1] In the second column, you find that 47.50% of the cases lie between the mean and this score. Since what is true to the right of the mean in a normal curve is also true to the left, we know that 95% (47.50 + 47.50 = 95.00) of the cases lie within this interval, as illustrated in Figure 18.3.

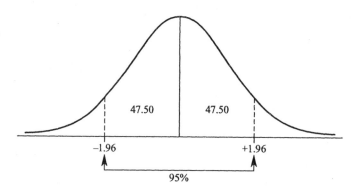

Figure 18.3. Percentage of cases between –1.96 and 1.96.

Table 1 applies to all normal curves. The table is abbreviated, showing only selected values of z because we will need only certain values for our work later in this book. More detailed tables can be found in statistics books available in most academic libraries should you have occasion to need to look up other values.

Term to Review Before Attempting Worksheet 18

table of the normal curve

---

[1]Note that 1.96 is one of the "values of special interest" at the end of the table. In Section 14, you learned that if you go out about 2 standard deviations on both sides of the mean, you capture *approximately* 95% of the cases. In Section 15, you learned you need to go out 1.96 standard deviations on both sides of the mean to capture precisely 95% of the cases.

Worksheet 18: Standard Scores and the Normal Curve

Riddle: How do many economics professors stay fit?

DIRECTIONS: To find the answer to the riddle, write the answer to each question in the space immediately below it. The word in parentheses in the solution section next to the answer to the first question is the first word in the answer to the riddle, the word beside the answer to the second question is the second word, and so on.

1. To two decimal places, what percentage of the cases in a normal distribution lies between the mean and a z-score of -1.00?

2. To two decimal places, what percentage of the cases in a normal distribution lies between the mean and a z-score of 1.96?

3. To two decimal places, what percentage of the cases in a normal distribution lies above a z-score of 1.00?

4. To two decimal places, what percentage of the cases in a normal distribution lies below a z-score of 3.00?

5. To two decimal places, what percentage of the cases in a normal distribution lies below a z-score of -2.00?

Worksheet 18 (Continued)

6. To two decimal places, what percentage of the cases in a normal distribution lies between a *z*-score of 1.00 and a *z*-score of −1.00?

7. To two decimal places, what percentage of the cases in a normal distribution lies between a *z*-score of 1.96 and a *z*-score of −1.96?

8. To two decimal places, what percentage of the cases in a normal distribution lies between a *z*-score of 2.58 and a *z*-score of −2.58?

Solution section:

50.00% (money)   68.26% (a)   99.02% (cycle)   100.00 % (exercise)

95.00% (business)   34.13% (they)   44.00% (jumping)   47.50% (ride)

15.87% (to)   0.13 (numbers)   99.87% (school)   2.28% (on)

49.51% (tenure)   84.13% (having)   99.51% (everything)

Write the answer to the riddle here, putting one word on each line: _____ _____ _____ _____ _____ _____ _____ _____

Section 19: Conceptual Introduction to Correlation

Correlation refers to the extent to which two variables are related across a group of subjects. Consider scores on the College Entrance Examination Board's *Scholastic Aptitude Test* (*SAT*) and first-year GPA in college. Because the *SAT* is widely used as a predictor in college student selection, there should be a correlation between these scores and GPA. Consider Example 1 in which *SAT-V* refers to the verbal portion of the *SAT*.[1] Is there a relationship?

Example 1:

Student	SAT-V	GPA
John	333	1.0
Janet	756	3.8
Thomas	444	1.9
Scotty	629	3.2
Diana	501	2.3
Hillary	245	0.4

Indeed, there is. Notice that students who scored high on the *SAT-V* (such as Janet and Scotty) had the highest GPAs. Also, those who scored low on the *SAT-V* (such as Hillary and John) had the lowest GPAs. This type of relationship is called **direct** or **positive**. In a direct relationship, those who score high on one variable tend to score high on the other, *and* those who score low on one variable tend to score low on the other.

Example 2 shows the relationship between a personality scale that measures willingness to take orders (on a scale from 0 to 20, where 20 represents eagerness to take orders) and the number of days served in county prison for a misdemeanor. Is there a relationship? If you consider the scores carefully, you will see that there is. Notice that those who have a high willingness to take orders (such as Jake and Dean) served the fewest days in prison, while prisoners

---

[1]*SAT-V* scores range from 200 to 800.

who have a low willingness (such as Jason and Joan) served the most days in prison. Such a relationship is called *inverse* or *negative*. In an inverse relationship, those who score high on one variable tend to score low on the other.

Example 2:

Prisoner	Willingness	Days Served
Jake	20	3
Jason	1	65
Sarah	10	20
Dick	11	24
Dean	18	7
Joan	3	40

It is important to note that just because we have established a correlation, we have not necessarily established a *causal relationship*. In a causal relationship, one variable is found to cause a change in the other—that is, one variable is found to affect the other. Consider the hypothetical relationship between willingness to take orders and days served in prison. Although a relationship was found, there could be many *causal* explanations. For example, those who have better attorneys may have had better instruction on how to be compliant (and, thus willing to take orders) and those same prisoners' attorneys, because they are better, may have had greater success in having them released from prison early. Many other explanations may be possible.

In order to determine *cause-and-effect*, a controlled *experiment* is needed in which different treatments are tried. For example, if a treatment given to an experimental group is shown to lead to a change not found in a comparable control group, then we would have evidence regarding causality. Note that in each of the two examples above, two variables were measured but no treatments were given. Even though we generally should not infer causality from a correlational study, we are still often interested in correlation. For example, the College Board and the colleges that use its test are interested in how well the test works in predicting success in college; it is not necessary to show what causes high GPAs in college in order to make the test useful. Also, correlations are of interest in developing theories. Often a postulate of a theory may say that X should be related to Y. If a correlation is found, it helps to support the theory.

Up to this point, we have examined the scores of only small numbers of subjects in clear-cut cases. However, in practice, large numbers of subjects are usually examined and there are, almost always, exceptions to the rule. Consider Example 3 in which we have the same students as in Example 1 but with the addition of two others—Joe and Patricia.

Example 3:

Student	SAT-V	GPA
John	333	1.0
Janet	756	3.8
Thomas	444	1.9
Scotty	629	3.2
Diana	501	2.3
Hillary	245	0.4
Joe	630	0.9
Patricia	404	3.1

Joe has a high *SAT-V* score but a very low GPA; he is an *exception* to the rule that high values on one variable are associated with high values on the other. There may be a variety of explanations for this exception—Joe may have had a family crisis during his first year in college *or* he may have abandoned his good work habits in favor of parties and booze as soon as he moved away from home to college. Patricia is another exception—perhaps she made an extra effort to apply herself to college work, which could not be predicted by the *SAT*. When studying hundreds of subjects, there will be many exceptions —some large and some small. To make sense of such data, some statistical techniques will be required, which will be explored in the next several sections.

Terms to Review Before Attempting Worksheet 19

correlation, direct or positive relationship,
inverse or negative relationship, causal relationship,
cause-and-effect, experiment

"I don't like to give a lot of homework over the weekend, so just read every other word."

Worksheet 19: Conceptual Introduction to Correlation

> *Riddle*: What happens just when you think you have hit bottom?

DIRECTIONS: To find the answer to the riddle, write the answer to each question in the space immediately below it. The word in parentheses in the solution section next to the answer to the first question is the first word in the answer to the riddle, the word beside the answer to the second question is the second word, and so on.

1. In the examples in this section, how many scores did each individual have?

2. What is another term for a *direct* relationship?

3. What is another term for an *inverse* relationship?

4. Do the scores in the box immediately below indicate a "direct" *or* an "inverse" relationship?

Student	Test A	Test B
Paula	9	3
Lucy	30	20
Mike	15	10
Christine	2	0
Rick	28	17

Worksheet 19 (Continued)

5. "Establishing a correlation establishes a causal relationship." Is this statement true or false?

6. "A controlled experiment is desirable in order to establish a cause-and-effect relationship." Is this statement true or false?

7. Do the scores in the box immediately below indicate a direct relationship?

Subject	Depression Scale	Cheerfulness Scale
Edward	80	50
John	90	40
Barbara	100	30
Cynthia	110	20
William	120	10

8. When working with hundreds of subjects, should an investigator expect many or few exceptions to an overall trend?

Solution section:

many (shovel)   few (low)   true (a)   false (you)   yes (illness)   2 (someone)

negative (above)   1 (loss)   inverse (weakness)   direct (tosses)   3 (sick)

positive (up)   no (bigger)   mixed (bourbon)   numerical (sickness)   4 (joining)

Worksheet 19 (Continued)

Write the answer to the riddle here, putting one word on each line: _____ _____ _____ _____ _____ _____ _____ _____

Notes:

Section 20: Scattergrams

A ***scattergram*** (also known as a ***scatter diagram*** or ***scatterplot***) is a graphic representation of the relationship between two variables. Here is the scattergram for the six subjects who took the *Scholastic Aptitude Test-Verbal* (*SAT-V*) and their GPAs earned during their freshman year in college. (See Example 1 in Section 19 for a listing of the scores.)

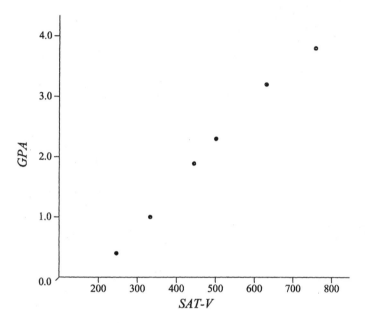

Figure 20.1. Scattergram for *SAT-V* scores and GPA.

There is one dot for each subject; the dot is placed where the two scores for that subject intersect. Thus, the dot in the upper right-hand corner is for Janet who had a *SAT-V* score of 756 and a GPA of 3.8. The dot in the lower left-hand corner is for Hillary who had a *SAT-V* score of 245 and a GPA of 0.4. The pattern is from the lower left-hand corner to the upper right-hand corner. This indicates that there is a direct (i.e., positive) relationship. Because the pattern is very clear, this relationship is characterized as *very strong*.[1]

---

[1] Notice that in terms of ranks, the relationship is perfect. That is, the subject with the highest *SAT-V* has the highest GPA, the subject with the second highest *SAT-V* has the next highest GPA, and so on, without exception.

111

These are guidelines for constructing scattergrams:

1. Draw two axes of *equal length*. Notice that GPAs can vary from 0.00 to only 4.00, whereas *SAT-V* scores can vary from 200 to 800. Yet, the two axes in Figure 20.1 are equal. This was done by planning the spacing for each in advance.

2. Use graph paper.

3. Label each axis with the name of a variable. If one variable was measured before the other (e.g., the College Board's *SAT* is administered before college GPAs are earned), it is customary to place the one measured first on the *x*-axis (i.e., the horizontal axis).

4. Place one dot where the two scores for each individual intersect.

5. Label the scattergram as a *figure* and give it a number and brief title.

Very strong relationships such as the one in Figure 20.1 are relatively rare in the social and behavioral sciences. Usually, there are a number of exceptions to the rule, and these exceptions create what is referred to as *scatter* on a scattergram. Figure 20.2 illustrates about the best we can expect when using a multiple-choice test to predict subsequent academic achievement.[2]

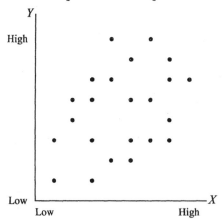

Figure 20.2. A direct relationship with many exceptions.

Notice that in Figure 20.2, the relationship is direct because the pattern of the dots is from lower-left to upper-right. There are, however, many

---

[2]In addition to the *SAT*, there are a variety of other types of tests designed to predict academic achievement such as algebra prognosis tests and reading readiness tests.

exceptions to the overall pattern from lower-left to upper-right. Because of these exceptions, the relationship is characterized as direct and only moderately strong.[3]

When a relationship is inverse, the pattern is from the upper-left hand corner to the lower right-hand corner, as illustrated in Figure 20.3. Notice that those subjects in the upper left-hand corner have low scores on Variable X but high scores on Variable Y. Those in the lower right-hand corner have high scores on Variable X but low scores on Variable Y.

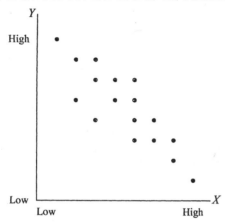

Figure 20.3. An inverse relationship.

Compare Figures 20.2 and 20.3. Not only are they different in direction (i.e., direct vs. inverse), they are also different in *strength*. The dots in Figure 20.3 cluster more tightly together than those in Figure 20.2. Therefore, Figure 20.3 represents a stronger relationship than that in Figure 20.2.

The three figures that we have examined so far all have a common feature; they illustrate what is known as ***linear relationships***. In a linear relationship, the dots form a pattern that follows a single straight line (even though there may be scatter around the line). Such relationships are frequently found.

Occasionally, investigators find ***curvilinear relationships***—one of which is illustrated in Figure 20.4. The dots in this figure form a curve that, starting from the left, goes up for a while (indicating a direct relationship), then turns downward (indicating an inverse relationship). Thus, the overall relationship is

---

[3]There are no specific scientific guidelines for labeling the strength of a relationship. What one person might call *strong*, another might call *moderate*.

neither direct nor inverse. Instead, it is described as *curvilinear*. Although curvilinear relationships are rare in the behavioral and social sciences, they can be extremely interesting. The variables for the hypothetical data in Figure 20.4 are anxiety and performance on a test of manual dexterity. The figure indicates that those who are extremely anxious about their performance and those who have very little anxiety both perform poorly on the test of dexterity. Those who are only moderately anxious perform best.

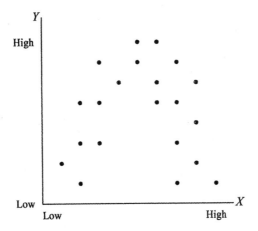

Figure 20.4. A curvilinear relationship.

Sometimes, there is no pattern in a scattergram; the dots are scattered everywhere, as in Figure 20.5. In this case, there is no discernible relationship.

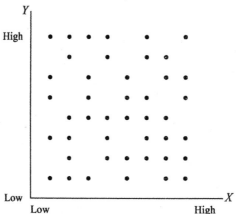

Figure 20.5. No relationship.

Constructing a scattergram should be the first step in analyzing the relationship between two variables. It gives the investigator an overview of the data and indicates whether it is linear. To describe any of the figures in this section with any accuracy would take at least a short paragraph. In the next several sections, you will learn of more concise ways to describe relationships.

Terms to Review Before Attempting Worksheet 20

scattergram (scatter diagram, scatterplot), linear relationships, curvilinear relationships

"The reason I'm successful in statistics class is because I'm lucky. But I didn't get lucky until I started working 90 hours a week!"

Worksheet 20: Scattergrams

DIRECTIONS: To find the answer to the riddle, write the answer to each question in the space immediately below it. The word in parentheses in the solution section next to the answer to the first question is the first word in the answer to the riddle, the word beside the answer to the second question is the second word, and so on.

1. How many scores for each person are required in order to draw a scattergram?

2. If the dots on a scattergram tend to go from the lower left-hand corner to the upper right, is the relationship "direct" *or* "inverse"?

3. If the dots on a scattergram tend to go from the upper left-hand corner to the lower right, is the relationship "direct" *or* "inverse"?

4. If there is no pattern formed by the dots on a scattergram (with dots scattered everywhere), is there a relationship?

5. What is the name for the type of relationship in which the dots follow a curve instead of a straight line?

Worksheet 20 (Continued)

6. What is the name for the type of relationship in which the dots follow a straight line?

7. Which one of the following scattergrams (A, B, or C) indicates the strongest relationship?

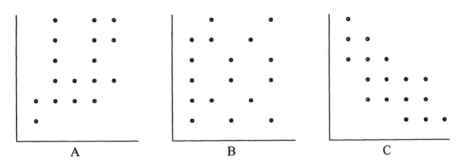

        A   B   C

8. Which one of the scattergrams (A, B, or C) indicates the weakest relationship?

Solution section:

1 (talking)	3 (self)	B (night)	C (every) curved (back)

1 (talking) 3 (self) B (night) C (every) curved (back)

linear (alone) none (lies) 2 (it) direct (gets) no (sitting)

A (fun) inverse (boring) lined (becoming) curvilinear (home) 4 (friends)

Write the answer to the riddle here, putting one word on each line: _____ _____ _____ _____

_____ _____ _____ _____

Notes:

Section 21: Introduction to the Pearson *r*

A statistician named Pearson developed a widely used statistic for describing the relationship between two variables. His statistic is often simply called the **Pearson r**. Its full, formal name is the *Pearson product-moment correlation coefficient,* and you might find variations on this in research literature such as *Pearson correlation coefficient* or the *product-moment correlation coefficient.*[1]

We'll begin by considering some basic properties of a *Pearson r*.

➔ It can range only from –1.00 to 1.00.

➔ –1.00 indicates a perfect negative relationship—the strongest possible inverse relationship.

➔ 1.00 indicates a perfect positive relationship—the strongest possible direct relationship.

➔ 0.00 indicates the complete absence of a relationship.

➔ The closer a value is to 0.00, the weaker the relationship.

➔ The closer a value is to –1.00 or 1.00, the stronger the relationship.

Thus, the *Pearson r* varies from –1.00 to 1.00, as illustrated here:

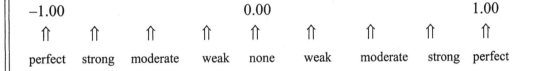

–1.00				0.00				1.00
⇑	⇑	⇑	⇑	⇑	⇑	⇑	⇑	⇑
perfect	strong	moderate	weak	none	weak	moderate	strong	perfect

Notice the labels *strong*, *moderate*, and *weak* are used in conjunction with both positive and negative values of *r*. Also, exact numerical values are not given for these labels. This is because the interpretation and labeling of an *r* may vary from one investigator to another and from one type of investigation to another. For example, one way to examine *test reliability* is to administer

---

[1]Pearson devised *r* for describing *linear relationships* (see Section 20). It should be used only for describing such relationships. When erroneously applied to a curvilinear relationship, it may indicate that there is little or no relationship when, in fact, there is a strong one.

the same test twice to a group of subjects without trying to change the subjects between administration of the tests. This will result in two scores per person, which can be correlated by calculating a Pearson *r*. In such a study, a professionally constructed test should yield high values of *r* such as .85 or higher.[2] A result such as .65 probably would be characterized as only moderately strong. In another type of study, where high values of *r* are seldom achieved (such as predicting college GPAs from College Board scores earned a year earlier), a value of .65 might be interpreted as strong or even very strong.[3]

The interpretation of the values of *r* is further complicated by the fact that an *r* is *not a proportion*. Thus, an *r* of .50 is not half of anything. It follows that multiplying .50 by 100 does *not* yield a percentage; that is, .50 is *not* equivalent to 50%. This is important because we are used to thinking of .50 as being halfway between 0.00 and 1.00. In Section 23, you will learn how to compute another statistic that is directly related to *r* but that may be interpreted as a proportion and converted to a percentage.

See Appendix B for additional information on interpreting a Pearson *r*.

Term to Review Before Attempting Worksheet 21

Pearson *r*

---

[2]A test is said to be reliable if its results are consistent. For example, if you measured the length of a table twice with a tape measure, you would expect very similar results both times—unless your measurement technique was unreliable.

[3]Careful study of the literature on the topic being investigated is needed in order to arrive at a non-numerical label or interpretation of a *Pearson r* that will be accepted by one's colleagues.

Worksheet 21: Introduction to the Pearson *r*

> *Riddle*: What is the wacky definition of a "baby-sitter"?

DIRECTIONS: To find the answer to the riddle, write the answer to each question in the space immediately below it. The word in parentheses in the solution section next to the answer to the first question is the first word in the answer to the riddle, the word beside the answer to the second question is the second word, and so on.

1. What letter of the alphabet is used to stand for Pearson's correlation coefficient?

2. "The closer a correlation coefficient is to –1.00, the weaker it is." Is this statement true or false?

3. "The closer a correlation coefficient is to 0.00, the weaker it is." Is this statement true or false?

4. What value of a correlation coefficient indicates a perfect, direct relationship?

5. Is a correlation coefficient of –.29 weaker than one of –.48?

6. Is a correlation coefficient of .68 stronger than one of –.89?

Worksheet 21 (Continued)

7. What value of a correlation coefficient indicates a perfect, inverse relationship?

8. What is the last name of the person who developed r?

Solution section:

b (baby)   0.00 (helps)   Pearson (television)   −1.00 (watch)   no (to)

r (a)   1.00 (gets)   100.00 (sleeping)   .68 (teenager)   true (who) c (cries)

false (person)   product (laughs)   10.00 (relies)   yes (paid)   .90 (expensive)

Write the answer to the riddle here, putting one word on each line: _____ _____ _____ _____ _____ _____ _____ _____

Section 22: Computation of the Pearson r

The original formula for r was defined in terms of standard scores (i.e., z-scores).[1] This formula is presented in Appendix C for those of you who wish to study it in order to better understand the meaning of r.

For those of you who are using calculators, this formula, known as the raw score or **computational formula for r**, is recommended:

$$r = \frac{N\Sigma XY - (\Sigma X)(\Sigma Y)}{\sqrt{[N\Sigma X^2 - (\Sigma X)^2][N\Sigma Y^2 - (\Sigma Y)^2]}}$$

To use the formula, you must first designate one set of scores as X and the other as Y. If one variable is measured before the other, it is customary to designate the first one measured as X and the other as Y. However, you will get the same answer if you reverse the designations when using this formula.

Table 22.1
Worktable for Computing r

Col. 1 Subject	Col. 2 X	Col. 3 Y	Col. 4 X^2	Col. 5 Y^2	Col. 6 XY
Bill	5	0	25	0	0
Liz	7	4	49	16	28
Carol	0	9	0	81	0
Stu	1	7	1	49	7
Frank	4	5	16	25	20
Brandy	2	6	4	36	12
$\Sigma =$	19	31	95	207	67

In Table 22.1, the scores are shown in columns 2 and 3. Begin your work by completing the worktable as follows: Compute the values in column 4 by squaring each value of X. Compute the values in column 5 by squaring each

---

[1]See Section 16 to review these.

value of *Y*. Finally, compute the values in column 6 by multiplying each *X* by each *Y*. (For example, Brandy's score of 2 on *X* is multiplied by her score of 6 on *Y* to obtain the product of 12 in column 6.) Then sum the values in columns 2 through 6.

The formula presented at the beginning of this section is rewritten here using column numbers to refer to the *sums* of the columns shown in Table 22.1.

$$r = \frac{N(Col.6) - (Col.2)(Col.3)}{\sqrt{[N(Col.4) - (Col.2)^2][N(Col.5) - (Col.3)^2]}}$$

$$r = \frac{6(67) - (19)(31)}{\sqrt{[(6)(95) - 19^2][(6)(207) - 31^2]}}$$

$$r = \frac{402 - 589}{\sqrt{[570 - 361][1242 - 961]}}$$

$$r = \frac{-187}{\sqrt{[209][281]}} = \frac{-187}{\sqrt{58729}} = \frac{-187}{242.341} = -.772 = -.77$$

The answer is negative, indicating that the relationship is inverse. With only a small number of scores, you can make a quick check to see whether the answer should be negative. In this case, subjects such as Carol and Stu had low scores on *X* and high scores on *Y*. Also, subjects such as Bill and Liz had high scores on *X* and low scores on *Y*. Remember that:

➔ in an *inverse relationship*, high scores on one variable are associated with low scores on the other, resulting in a negative value of *r*.

➔ in a *direct relationship*, high scores are associated with high scores *and*

  low scores are associated with low scores, resulting in a positive value of *r*.

As another partial check, keep in mind that if the value of *r* is going to be negative, it is the numerator of the fraction (e.g., −187) that will be negative; the denominator is never negative. If you obtain a negative in the denominator, you know you have made a mistake in your work.

When subjects are *ranked* on both variables (e.g., the most talented is given a rank of 1, the next most talented is given a rank of 2, etc.), then a simpler formula developed by Spearman may be applied. This formula is presented in Appendix D.

Terms to Review Before Attempting Worksheet 22

computational formula for *r*, inverse relationship, direct relationship

"How many statisticians does it take to screw in a light bulb? Three: One to analyze the data, one to draw conclusions, and one to *skew* it in."

Worksheet 22: Computation of the Pearson *r*

Riddle: In what footsteps do most boys follow?

DIRECTIONS: To find the answer to the riddle, write the answer to each question in the space immediately below it. The word in parentheses in the solution section next to the answer to the first question is the first word in the answer to the riddle, the word beside the answer to the second question is the second word, and so on.

Questions 1 through 5 refer to these data.

Col. 1 Subject	Col. 2 X	Col. 3 Y
Victor	2	0
Juliet	6	7
Romeo	4	5
Millie	3	6
Rob	5	4

1. What is the value of N?

2. What is the value of the sum of column 2?

3. What is the value of $\sum X^2$?

4. What is the value of $\sum XY$?

Worksheet 22 (Continued)

5. What is the value of r?

Questions 6 through 9 refer to these data.

Col. 1 Subject	Col. 2 X	Col. 3 Y
Ginny	12	1
Leslie	10	2
Steve	7	4
Clyde	8	3
Dave	6	6
Jose	5	5

6. What is the value of $\sum Y$?

7. What is the value of $\sum Y^2$?

8. What is the value of $\sum XY$?

9. What is the value of r?

Worksheet 22 (Continued)

Solution section:

> 5 (those)   100 (father)   .94 (shoes)   145 (covered)   .75 (path)
>
> −.94 (up)   91 (had)   89 (becoming)   25 (listening)   21 (he)
>
> 125 (watches)   20 (that)   .70 (thought)   −.64 (never)   90 (his)

Write the answer to the riddle here, putting one word on each line: _____ _____ _____ _____ _____ _____ _____ _____ _____

Section 23: Coefficient of Determination

The ***coefficient of determination*** is useful when interpreting a Pearson r. Its symbol, r^2, explains how it is computed; to obtain it, simply square r. Thus, for a Pearson r of .60, r^2 equals .36 (.60 × .60 = .36).

Although the computation is simple, what it indicates is sometimes difficult for students to grasp. Let's begin with a specific example from Section 22, in which these scores were shown:

Table 23.1
Scores from Section 22

Col. 1	Col. 2	Col. 3
Subject	X	Y
Bill	5	0
Liz	7	4
Carol	0	9
Stu	1	7
Frank	4	5
Brandy	2	6

Notice that there are differences among the scores on variable X; this is referred to as ***variance***. There is also variance in the scores on variable Y. When interpreting a Pearson r, an important question is: *What percentage of the variance on one variable is accounted for by the variance on the other?* If we are trying to predict variable Y (which might be college GPAs) from variable X (which might be scores on a college admissions test), the question might be phrased as: *What percentage of the variance on Y is <u>predicted</u> by the variance on variable X?* The answer to the question is determined simply by determining r^2 and multiplying it by 100. For the scores shown in Table 23.1, $r = -.77$. Thus,

$$-.77 \times -.77 = .59 \times 100 = 59\%$$

We have determined that 59% (*not* 77%) of the variance on one variable is accounted for by the variance on the other in this example.[1]

Let's put the 59% in perspective. Suppose we are trying to predict how subjects will score on variable Y. Suppose we naively put all of the subjects' names on slips of paper in a hat and draw a name and declare that the first name drawn will probably perform best on variable Y, and draw a second name and declare that this person will probably perform second best on variable Y, and so on. What percentage of the variance on Y will we predict using this procedure? In the long run, with large numbers of subjects, the answer is about zero (0.00) percent. In the above example, variable X accounted for 59% of the variance on variable Y, which is substantially better than using a random process to make predictions.

It follows, however, that if we can account for 59% of the variance, 41% (100% − 59% = 41%) of the variance is *not* accounted for. Thus, there is much room for improvement in our ability to predict.[2]

Table 23.2
Pearson r and Related Statistics

r	r^2	% accounted for	% *not* accounted for
.10	.01	1%	99%
.20	.04	4%	96%
.30	.09	9%	91%
.40	.16	16%	84%
.50	.25	25%	75%
.60	.36	36%	64%
.70	.49	49%	51%
.80	.64	64%	36%
.90	.81	81%	19%
1.00	1.00	100%	0%

[1] *Variance accounted for* is sometimes referred to as *explained variance*.
[2] Notice that because r is negative (i.e., −.77), we need to predict that those who score high on X will score low on Y and that those who score low on X will score high on Y.

Consider Table 23.2. It shows selected values of r, the corresponding values of r^2, and the percentage of variance accounted for and not accounted for. Notice that small values of r shrink dramatically when converted to r^2, indicating that we should be very cautious when interpreting small values of r—they are further from perfection than they might seem at first.[3]

Keeping Table 23.2 in mind when reading scholarly articles should give you pause because many authors interpret an r in the .20 to .40 range as being important. In fact, they may be of some practical importance under certain circumstances. However, keep in mind that in this range, 84% to 96% of the variance on one variable is *not* accounted for by the other. When considering a prediction study, an r of .40 leaves much room for improvement when attempting to predict variable Y using variable X.

In the next section, you will learn how we can often improve our ability to predict by using two predictor variables when making predictions.

Terms to Review Before Attempting Worksheet 23

coefficient of determination, r^2, variance

---

[3]Note that when there is no variance on either variable, the Pearson r will equal 0.00 and r^2 will also equal 0.00. This is easy to see by example. For instance, suppose a group of students all had identical *SAT* scores. Since these scores fail to differentiate among students, they cannot predict who will have high GPAs, who will have average GPAs, etc.

Worksheet 23: Coefficient of Determination

Riddle: **Why is the government certain that there is no life on Mars?**

DIRECTIONS: To find the answer to the riddle, write the answer to each question in the space immediately below it. The word in parentheses in the solution section next to the answer to the first question is the first word in the answer to the riddle, the word beside the answer to the second question is the second word, and so on.

1. "For an r of .55, 55% of the variance on one variable is accounted for by the variance on the other." Is this statement true or false?

2. If $r = .24$, what is the value of r^2?

3. If $r = .47$, what is the value of the coefficient of determination?

4. Is r or r^2 a more direct indicator of variance accounted for?

5. If $r = .88$, what percentage of the variance on one variable is accounted for by the variance on the other?

6. If $r = .66$, is the majority of the variance on one variable accounted for by the variance on the other?

Worksheet 23 (Continued)

7. If $r = .78$, is the majority of the variance on one variable accounted for by the variance on the other?

8. If $r = .95$, what percentage of the variance on one variable is *not* accounted for by the variance on the other?

9. If $r = .25$, what percentage of the variance on one variable is *not* accounted for by the variance on the other?

Solution section:

94% (money)	90% (shuttle)	6% (walking)	10% (any)
no (States)	true (astronaut) yes (for)	r (exploration)	.88 (not)
false (they)	r^2 (the) .22 (asked)	.48 (green)	77% (United)
.06 (haven't)	25% (overwhelmed)	95% (unconventional)	

Write the answer to the riddle here, putting one word on each line: _____ _____ _____ _____
_____ _____ _____ _____ _____

Notes:

Section 24: Multiple Correlation

Often, we are interested in the extent to which two variables, in combination, predict a third variable. For example, we may wish to know how well high school GPAs in combination with scores on a college admissions test predict college GPAs. In this example, we have three variables and, thus, three scores per subject:

Variable 1: college GPAs

Variable 2: high school GPAs

Variable 3: scores on a college admissions test

In order to use the formula presented below, you must name the variable you are trying to predict as Variable 1; it does not matter which of the others you name Variable 2 and which you name Variable 3.

Because there are three variables, three values of Pearson r should be computed, which will be identified with subscripts. For example, r_{12} stands for the relationship between Variables 1 and 2. Consider this data:

$$r_{12} = .55$$
$$r_{13} = .44$$
$$r_{23} = .38$$

The value of r_{12} indicates how well high school GPAs predict college GPAs, and the value of r_{13} indicates how well scores on the admissions test predict college GPAs. Clearly, high school GPA is a better predictor than the admissions test. Finally, the value of r_{23} indicates the extent to which the two predictors are correlated; the .38 indicates that, to a modest extent, there is overlap between the two predictors.

Getting back to our original concern, we want to know the extent to which Variables 2 and 3, *in combination*, will predict Variable 1.[1] To answer the question, we compute a ***multiple correlation coefficient***, whose symbol is ***R***. The formula for R when the variable being predicted has been named Variable 1 is:

$$R = \sqrt{\frac{r_{12}^2 + r_{13}^2 - 2r_{12}r_{13}r_{23}}{1 - r_{23}^2}}$$

$$R = \sqrt{\frac{.55^2 + .44^2 - (2)(.55)(.44)(.38)}{1 - .38^2}}$$

$$R = \sqrt{\frac{.303 + .194 - .184}{1 - .144}}$$

$$R = \sqrt{\frac{.313}{.856}} = \sqrt{.366} = .605 = .60$$

For all practical purposes, the multiple R is interpreted in the same way as the Pearson r except that, in this case, R indicates how well two variables in combination predict another one.[2] Note that the variable being predicted is often called the ***criterion variable***.

The example shown above illustrates an interesting point. Notice that if we just use high school GPAs to predict college GPAs, the Pearson r is .55. If we just use admissions test scores to predict college GPAs, the Pearson r is .44. Together, in combination, the two predictors bring us up only to a multiple R of .60. At first, you might think that the .55 and .44 in combination might bring us close to 1.00 (keeping in mind that correlation coefficients cannot exceed 1.00). The relatively modest increase in our ability to predict is due to the fact that the two predictors overlap—that is, they both tap some of the

---

[1]The formula presented here tells us how well the combination will work if combined in the best mathematical fashion. The mathematics of determining such a combination is beyond the scope of this book.

[2]The technique of multiple correlation can be expanded to include more than two predictors. However, the mathematics for doing this are cumbersome and beyond the scope of this book.

same skills, as indicated by the *r* of .38 for the relationship between the two predictors. In general, the greater the correlation between the two predictors, the less increase we get when we use them in combination.

Terms to Review Before Attempting Worksheet 24

multiple correlation coefficient, *R*, criterion variable

"On average, I feel fine."

Worksheet 24: Multiple Correlation

> *Riddle*: What should you do on the "keyboard of life"?

DIRECTIONS: To find the answer to the riddle, write the answer to each question in the space immediately below it. The word in parentheses in the solution section next to the answer to the first question is the first word in the answer to the riddle, the word beside the answer to the second question is the second word, and so on.

1. How many scores must you have for each subject in order to compute a multiple correlation coefficient using the formula given in this section? (In other words, how many variables must you have?)

2. To use the formula for computing the multiple correlation coefficient presented in this section, should the variable being predicted be named number 1, number 2, or number 3?

3. What is the symbol for the multiple correlation coefficient?

4. "As defined in this section, r_{23} indicates the extent to which the best predictor variable is correlated with the criterion variable." Is this statement true or false?

5. "The formula for the multiple correlation coefficient presented in this section indicates the extent to which two variables in combination predict a criterion variable." Is this statement true or false?

138

Worksheet 24 (Continued)

6. Is a *criterion variable* a predictor variable or the variable being predicted?

7. What is the value of the multiple correlation coefficient given these Pearson rs:
$r_{12} = .30, r_{13} = .20, r_{23} = .10$?

8. What is the value of the multiple correlation coefficient given these Pearson rs:
$r_{12} = .61, r_{13} = .60, r_{23} = .47$?

Solution section:

.34 (escape)	number 2 (labor)	.12 (shaking)	.71 (key)	true (on)
variable being predicted (the)	number 3 (pain)	r (crying)	.50 (small)	
3 (always)	R (one)	predictor variable (visible)	false (finger)	
number 1 (keep)	.30 (hurt)	r_{23} (spank)	.10 (doctor)	.47 (hospital)

Write the answer to the riddle here, putting one word on each line: _____ _____ _____ _____
_____ _____ _____ _____

Notes:

Section 25: Introduction to Linear Regression

In the previous several sections, we have been considering relationships *across groups of subjects*—that is, to what extent variables are correlated when the scores of groups are examined.

Frequently, when we know that there is a moderate to strong correlation across a group of subjects, we want to make predictions for new individuals who are subsequently tested. For example, suppose that last year we administered an algebra aptitude test to a group of subjects during the first week of an algebra course; the test measured basic math skills needed to learn algebra. Then, we administered an algebra achievement test at the end of the course and constructed the scattergram shown in Figure 25.1.[1]

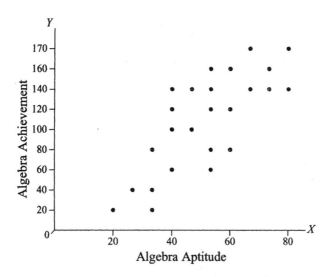

Figure 25.1. Scattergram.

How could we use the information in the scattergram to make predictions for students enrolling in algebra this year? We could make rough predictions, of course. For example, if Frank obtains an algebra aptitude score of 35, we

---

[1]To review scattergrams, see Section 20.

can see from the scattergram that he probably would not do well since his aptitude score puts him in the lower left-hand corner. If Sarah obtains a score of 75 on aptitude, we could predict that she probably will do well on achievement because those who scored about 75 last year obtained high achievement test scores.

We can be more specific in our predictions if we use the technique of **linear regression**. In this technique, we determine the equation for a single straight line that best describes the dots.[2] Then, when we make predictions, we use the equation to get a single predicted score.

Let's first review the general equation for any straight line. It is:

$$Y = a + bX$$

Where:

Y is the score on variable Y (the score to be predicted)[3]

a is the **intercept** (the point where the straight line meets the y-axis)

b is the **slope** (this determines the angle of the line)

X is the score on variable X

Figure 25.2 shows a solid line with an intercept (a) of 2 and a slope (b) of .5 units. The intercept is easy to see; it is the point at which the line meets the y-axis. The slope indicates the ratio of change that produces the angle of the line. In this figure, the dashed lines show that for every one unit increase on X (whether it's an inch or a mile!), we need to go up .5 units on Y to reach the solid line. Put another way, the vertical dashed line is one-half (i.e., .5) the length of the horizontal line. Thus, the formula for the solid line in Figure 25.2 is $Y = 2 + .5X$. Given the intercept and slope, we can make predictions for individuals once we know their score on X. Here are two predictions:

---

[2]Obviously, no single straight line can go through all the dots when there is scatter on the scattergram. The *best line* is defined as the one that minimizes the squared differences of the dots from the line.

[3]The symbol Y' (pronounced "Y prime") is sometimes used to stand for the predicted score.

John has a score on X of 14. His predicted score on Y is:

$$Y = 2 + (.5)(14) = 2 + 7 = 9$$

Joan has a score on X of 6. Her predicted score on Y is:

$$Y = 2 + (.5)(6) = 2 + 3 = 5$$

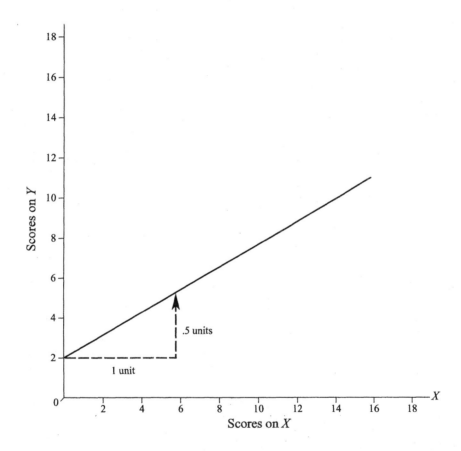

Figure 25.2. Line with a positive slope.

Figure 25.3 shows a line with a negative slope. Because the intercept is 12 and the slope is –1.5, the formula is $Y = 12 + -1.5X$. Notice that because the slope is negative, when we go right one unit, we go *down* 1.5 units.

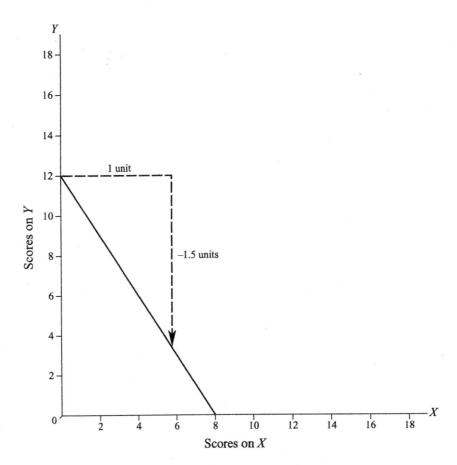

Figure 25.3. Line with a negative slope.

The following is a prediction based on the line in Figure 25.3:

Juliet has a score on X of 4. Her predicted score on Y is:

$$Y = 12 + (-1.5)(4) = 12 + -6 = 6$$

Keep in mind that linear regression is useful only if the dots form a pattern that follows a straight line (i.e., a linear relationship—*not* a curvilinear one).[4]

In the next section, you will learn how to compute the values of *a* and *b* for a set of data.

---

[4]See Section 20 to review the distinction between *linear* and *curvilinear* relationships.

Terms to Review Before Attempting Worksheet 25

linear regression, intercept, slope

Allen gets a love letter from the president of the Statistics Club.

Worksheet 25: Introduction to Linear Regression

> **Riddle:** If you do a good deed, why should you get a receipt?

DIRECTIONS: To find the answer to the riddle, write the answer to each question in the space immediately below it. The word in parentheses in the solution section next to the answer to the first question is the first word in the answer to the riddle, the word beside the answer to the second question is the second word, and so on.

1. Is X or Y defined as the score to be predicted?

2. Is the point where the line meets the y-axis called the "slope" *or* the "intercept"?

3. "The purpose of linear regression is to determine the direction and strength of relationships between two variables across a group of subjects." Is this statement true or false?

4. In $Y = a + bX$, is a or b the slope?

5. If a line has a negative slope, will the line slope up or down?

6. If $Y = 3.1 + 1.6X$ and if Richard has a score of 12 on X, what is his predicted score on Y?

Worksheet 25 (Continued)

7. If $Y = 50.2 + -2.4X$ and if Anne has a score of 20 on X, what is her predicted score on Y?

8. Is the technique described in this section appropriate for use with curvilinear relationships?

9. "In general, the lower the correlation between two variables, the greater the error that will be made when using linear regression." Is this statement true or false?

Solution section:

56.40 (charity)   no (revenue)   true (service)   −2.20 (external)   2.20 (internal)

98.20 (like)   yes (home) Y (in) X (visible) a (helpful)

22.30 (the)   down (like)   up (lift)   slope (silly)   intercept (case)

false (heaven) b (is)   47.80 (helpless)   4.70 (people)

Write the answer to the riddle here, putting one word on each line: _____ _____ _____ _____
_____ _____ _____ _____ _____

Notes:

Section 26: Computations for Linear Regression

The formulas for calculating linear regression will be illustrated using the scores from Table 22.1 in Section 22, which is reproduced here as Table 26.1.

Table 26.1
Worktable for Computing Slope and Intercept

Col. 1	Col. 2	Col. 3	Col. 4	Col. 5	Col. 6
Subject	X	Y	X^2	Y^2	XY
Bill	5	0	25	0	0
Liz	7	4	49	16	28
Carol	0	9	0	81	0
Stu	1	7	1	49	7
Frank	4	5	16	25	20
Brandy	2	6	4	36	12
$\Sigma =$	19	31	95	207	67

The correlation coefficient for the relationship between scores on X and Y, calculated to be $-.77$ in Section 22, indicates that those who are high on X tend to be low on Y and those who are low on X tend to be high on Y. If we wish to predict scores on variable Y from scores on variable X, we need to determine the equation for the straight line that maximizes the predictability of the scores.[1] As you know from the previous section, this means that we need to calculate the **slope** (**b**) and the **intercept** (**a**).

To calculate the slope (*b*), use this formula:[2]

$$b = \frac{\Sigma XY - [(\Sigma X)(\Sigma Y)/N]}{\Sigma X^2 - [(\Sigma X)^2/N]}$$

---

[1]A precise estimate of the degree of accuracy in prediction can be made with the standard error of estimate, which is discussed in Appendix E. The formulas in this section define a line that minimizes the squared distance of the scores from the line.

[2]Review Section 22 on computation of the correlation coefficient if you need to review how to set up the worktable that produces the values needed for the formulas in this section. Note that the Y^2 column, which is needed to calculate a correlation coefficient, is not needed for linear regression.

$$= \frac{67 - [(19)(31)/6]}{95 - [19^2/6]} = \frac{67 - [589/6]}{95 - [361/6]} = \frac{67 - 98.167}{95 - 60.167} = \frac{-31.167}{34.833} =$$

$$-.8947 = -.89$$

The next step is to calculate the means of X and Y, as follows:

$$\text{Mean of } X (M_x) = 19/6 = 3.17$$
$$\text{Mean of } Y (M_y) = 31/6 = 5.17$$

Third, calculate the intercept (a) by substituting the values of the slope and the means into this formula:

$$a = M_y - bM_x$$

$$a = 5.17 - (-.89)(3.17) = 5.17 - (-2.82) = 7.99 = 8.00$$

Because the general formula for a straight line is:

$$Y = a + bX,$$

the equation for the data in question is:

$$Y = 8.00 + -.89X$$

Now we have an equation into which we can substitute a value of X for a new individual and obtain a predicted value of Y. For example, suppose we test Robert using test X, and he obtains a score of 6. His predicted score on Y would be calculated as follows:

$$Y = 8.00 + (-.89)(6) = 8.00 + -5.34 = 2.66$$

In other words, we can use a group's performance on two variables in the past to predict new individuals' performances in the future.

You can make a rough check of your work by making an estimate, as we did for Robert. Notice that he obtained a score of 6, which, in Table 26.1, is near the top of the group on variable X. Does it make sense that his predicted score is 2.66? Let's look at two subjects who scored near him on X. The scores Bill and Liz earned on X and Y are in Table 26.2.

Table 26.2
The Two Highest Scoring Subjects on Test X from Table 22.1 Plus Robert, Who Had a High Score on Test X

Subject	Test X	Test Y
Bill	5	0
Robert	*6*	*2.66*
Liz	7	4

Since Robert's score on X is between Bill and Liz's scores, it makes sense that the predicted score for Robert should be between Bill and Liz's scores on Y, as it is.

When there are a large number of scores, a better way to make a rough check on your work is to draw a scattergram for the relationship (see Section 20) and then draw the line, using the slope and intercept that you calculated. If there is a clear trend in the pattern created by the dots, the line should follow that trend (that is, there should be some scatter on both sides of the line).

Terms to Review Before Attempting Worksheet 26

slope, intercept

Worksheet 26: Computations for Linear Regression

> ### *Riddle*: How do marriage and celibacy differ?

DIRECTIONS: To find the answer to the riddle, write the answer to each question in the space immediately below it. The word in parentheses in the solution section next to the answer to the first question is the first word in the answer to the riddle, the word beside the answer to the second question is the second word, and so on.

The questions are based on the following table.

Table 26.3

Worktable for Computing Slope and Intercept

Col. 1 Subject	Col. 2 X	Col. 3 Y	Col. 4 X^2	Col. 5 Y^2	Col. 6 XY
Carol	0	2			
Wayne	1	4			
Paula	4	5			
Bobby	3	7			
David	7	8			
$\Sigma =$					

1. What is the value of ΣX?

2. What is the value of ΣY?

3. What is the value of ΣX^2?

4. What is the value of ΣXY?

Worksheet 26 (Continued)

5. What is the value of *b* (the slope)?

6. What is the value of the mean of *Y*?

7. What is the value of *a* (the intercept)?

8. If a new person obtains a score of 8 on test *X*, what is the predicted value of *Y* for that person?

9. If a new person obtains a score of 2 on test *X*, what is the predicted value of *Y* for that person?

Solution section:

75 (many)	101 (thorns)	3.000 (peacefully)	−29.810 (themselves)	4.433 (roses)
15 (marriage)	26 (has)	9.035 (no)	−4.36 (afterwards)	158 (loving)
103 (funny)	16 (wary)	.767 (but)	5.200 (celibacy)	2.899 (has)

Write the answer to the riddle here, putting one word on each line: _____ _____ _____ _____

_____ _____ _____ _____ _____

Notes:

Section 27: Introduction to Sampling

Inferential statistics help us make inferences (i.e., generalizations) from samples to populations. For example, we might be interested in the attitudes of all registered nurses in Maryland toward people with AIDS. The nurses would constitute our population. If we administered an AIDS attitude scale to all these nurses, we would be studying the population, and the summarized results (such as means and standard deviations) would be referred to as *parameters*. If we studied only a sample of the nurses, the summarized results would be referred to as *statistics*.

No matter how we sample, it is always possible that the *statistics* we obtain do not accurately reflect the *population parameters* that we would have obtained if we had studied the entire population. In fact, we always expect some amount of error when we have sampled.

If sampling creates errors, why do we sample? First, it is not always possible for economic and physical reasons to examine an entire population. Second, with proper sampling, we can obtain highly reliable results for which we can estimate the amount of error to allow for in our interpretations of the data.

The most important characteristic of a good sample is that it be free from *bias*. A bias exists whenever some members of a population have a greater chance of being selected for inclusion in the sample than others. Here are some examples of biased samples:

➜ A professor wishes to study the attitudes of all sophomores at a college (the population) but asks only those enrolled in her introductory psychology class (the sample) to participate in the study. Note that only those in the class have a chance of being selected; all other sophomores have no chance.

➜ A person wishes to predict the results of a citywide election (the population) but asks the intentions of only voters whom he encounters in a large shopping mall (the sample). Note that only

155

those in the mall whom he decides to approach have a chance of being selected; all other voters have no chance.

➜ A magazine editor wishes to determine the opinions of rifle owners on a gun-control measure (the population) but mails questionnaires only to those who subscribe to her magazine, which appeals to sporting enthusiasts (the sample). Note that only subscribers have a chance to respond; all other rifle owners have no chance.

In the examples above, *samples of convenience* (or *accidental samples*) were used, increasing the odds that some will be selected and reducing the odds that others will—but there is an additional problem. Even those who have a chance of being included may refuse to participate. This problem is often referred to as *volunteerism*. Volunteerism is presumed to create a bias because those who decide not to participate have no chance of being included. Furthermore, many studies comparing participants with nonparticipants suggest that participants tend to be more highly educated and from higher socioeconomic status (SES) groups than their counterparts. Efforts to reduce the effects of volunteerism include offering rewards, stressing to potential participants the importance of the study, and making it easy for people to respond such as providing them with a self-addressed, stamped envelope.

To eliminate bias, some form of *random sampling* is needed. A classic form of random sampling is *simple random sampling*.[1] In this technique, each member of the population is given an equal chance of being selected. A simple way to accomplish this with a small population is to put the names of all members of a population on slips of paper, thoroughly mix the slips, and have a blindfolded assistant select the number desired for the sample. After the names have been selected, efforts must be made to encourage all of those selected to participate. If some members refuse, as often happens, we have a biased

---

[1]Another method for selecting a *simple random sample* and other types of random samples are described in the next section.

sample even though we started by giving everyone an equal chance to have his or her name selected.

But let's suppose that we are fortunate. We have selected names using simple random sampling and we have obtained the cooperation of everyone selected. In this case, we have an *unbiased sample*. Can we be certain that the results we obtain from the sample accurately reflect those we would have obtained by studying the population? Certainly not! We now have the possibility of random errors, which statisticians simply call *sampling errors*. At random (i.e., by chance), we may have selected a disproportionately large number of Democrats, males, low SES group members, etc. Such errors may affect our results.

If both biased and unbiased sampling are subject to error, why do we prefer unbiased sampling? We prefer it for two related reasons: (1) Inferential statistics allow us to estimate the amount of error to allow for when analyzing the results from unbiased samples, and (2) the amount of error obtained from unbiased samples is small when large samples are used.

It is important to note that selecting a very large biased sample does not reduce potential errors. For example, if the person who is trying to predict the results of a citywide election is very persistent and spends weeks at the shopping mall asking all registered voters that he encounters how they intend to vote, he will obtain a very large sample of people who may differ from the population of voters in various ways—such as being more affluent, having more time to spend shopping, etc. Increasing the sample size, in this case, has not reduced the amount of error due to bias.

Suppose that we have wisely decided to use random sampling. How large should our sample be? This depends on a variety of factors, but here are a few generalizations that guide researchers:

1. The larger the sample, the better—but increasing sample size produces diminishing returns. For example, using a sample of 200 instead of 100 has a much greater effect on reducing sampling errors than using a sample of 3,100 instead of 3,000. In

concrete terms, this means that an extra 100 subjects added to a small sample has a much greater effect on precision than adding 100 subjects to a large sample. In fact, in many national surveys, samples of only 1,500 to 2,000 carefully selected subjects yield highly accurate results.

2. When there is little variability in a population, even a small sample may yield highly accurate results. For example, if you take a random sample of eggs that have been graded as "extra large" and weigh them, you will probably find only a small amount of variation among them. For this population, a small random sample should yield an accurate estimate of the average weight of "extra large" eggs.

3. When there is much variability in a population, small samples may produce data with much error. Suppose you wanted to estimate the math achievement of sixth graders in a very large metropolitan school district and drew a random sample of only 100 students. Because there is likely to be tremendous variation in math ability across a large school district, a sample of 100, even though it is random, could be very misleading. By chance, for example, you may obtain a disproportionately large number of high achievers. Using a much larger sample would greatly reduce this possibility.

There are times, however, when it is impractical or impossible to use a random sample, but information is needed from a sample in order to make a decision. For example, a soft-drink manufacturer wishes to test market a new soda knowing it is impractical to obtain a random sample of all potential customers in the United States. In this case, the manufacturer might resort to some form of *quota sampling*. Using previously collected national statistics on what the typical soda drinker is like in terms of gender, ethnicity, and income, the manufacturer might test the new product with a sample that has the same gender, ethnic, and income characteristics—but obtain the sample by screening people at public places or by going door-to-door until enough men and enough women are found, enough people from each ethnic group are found, etc. Notice that this technique is *not* random and is subject to bias. Even if the manufacturer has the correct proportion of males, for example, the males selected might tend to be from a particular region of the United States that has different tastes than the national population of male soda drinkers. Thus, the results of quota sampling should be viewed with skepticism.[2]

Strictly speaking, the inferential statistics in the rest of this book should be applied only to data obtained from random samples. In practice, however, they are often misapplied. When you encounter them applied to results based on biased samples in journals and other sources, they should be regarded as only *very rough* rules of thumb.

I do not mean to leave you with the impression that all research in which biased samples are used produce worthless results. There are many situations in which researchers have no choice but to use biased samples. For example, for ethical and legal reasons, much medical research is done on volunteers who are willing to take the risk of using a new medication or undergoing a new procedure. If promising results are obtained in initial studies, larger studies with better (but usually still biased) samples are undertaken. At some point, despite the possible role of bias, decisions—such as Food and Drug Administration approval of a new drug—need to be made on the basis of biased samples. Little progress would be made in most fields if the results of all studies with biased samples were summarily dismissed.

---

[2]Another approach in this situation is to test market the product in selected areas of the country, using areas that are believed to be *typical* of the country at large. This is called *purposive sampling* because the collection of areas are believed to have residents with characteristics similar to those of the national population.

Terms to Review Before Attempting Worksheet 27

parameters, statistics, bias, samples of convenience (accidental samples), volunteerism, random sampling, simple random sampling, unbiased sample, sampling errors, quota sampling

"My answers aren't wrong.
They just have random features."

Worksheet 27: Introduction to Sampling

Riddle: Why are people who can laugh at themselves fortunate?

DIRECTIONS: To find the answer to the riddle, write the answer to each question in the space immediately below it. The word in parentheses in the solution section next to the answer to the first question is the first word in the answer to the riddle, the word beside the answer to the second question is the second word, and so on.

1. Suppose the population of first-grade teachers in a school district was surveyed on their attitudes toward a curriculum change, and all of the teachers in the population participated. Then percentages were computed. Are the percentages "statistics" *or* "parameters"?

2. A researcher wishes to generalize to all sociology students and uses volunteers in a sociology class as subjects. Is the sample "biased" *or* "unbiased"?

3. Suppose the names of all members of a population are put on slips of paper, the slips are thoroughly mixed, and 10% of the names are drawn for the sample. Is the group selected for the sample "biased" *or* "unbiased"?

4. What is the name of the type of sampling described in question 3?

5. Can we be certain that there are no sampling errors if we use random sampling?

Worksheet 27 (Continued)

6. "Using a very large sample is an effective way to reduce the errors created by a bias in sampling." Is this statement true or false?

7. Assume that Population A has very little variability and that Population B has much variability. For which population would a larger sample be needed to get precise results?

8. "Adding individuals produces diminishing returns in terms of reducing sampling errors." Is this statement true or false?

Solution section:

```
    parameters (because)   statistics (happy)   quota (are)   biased (they)

        yes (becoming)   unbiased (will)   Population A (thoughtful)

   simple random sampling (never)   sample size (joke)   no (cease)   false (to)

 true (amused)   Population B (be)   sample size (speed)   volunteerism (laughter)
```

Write the answer to the riddle here, putting one word on each line: _____ _____ _____ _____ _____ _____ _____ _____

Section 28: A Closer Look at Sampling

As noted in the previous section, an unbiased sample may be obtained by using *simple random sampling*. Putting names on slips of paper and drawing the number needed for the sample is a classic method of obtaining such a sample. For larger populations, it is more efficient to use a *table of random numbers*, a portion of which is shown in Table 2 near the end of this book.[1] In this table, there is no sequence to the numbers and, in a large table, each number appears about the same number of times. To use the table, first assign everyone in the population a *number name*. For example, if there are 90 people in the population, name the first person 01, the second person 02, the third person 03, etc., until you reach the last person whose number is 90.[2] (Computerized records have the individuals already numbered, which simplifies the process; any set of numbers will work as number names as long as each one has a different number and each one has the same number of digits in his or her name.) To use the table, flip to any page in a book of random numbers and put your finger on the page without looking; this will determine your starting point. We will start in the upper-left-hand corner of Table 2 for the sake of illustration. Because each person has a two-digit number name, the first two digits identify our first subject; this is person number 21. The next two digits to the right (ignoring the spaces between the columns, which are provided only as a visual aid while reading the table) are 0 and 4; thus, person number 04 will also be included in our sample. The third number is 98. Because there are only 90 people in the population, skip 98 and continue to the right to 08, which is the number of the next person drawn. Continue moving across the rows to select the sample.

Stratified random sampling is usually superior to *simple random sampling*. In this technique, the population is first divided into strata that are believed to be relevant to the variable(s) being studied. Suppose, for example,

---

[1]Academic libraries have books of random numbers. Statistical computer programs can also generate them.

[2]The number of digits in the number names must equal the number of digits in the population total. For example, if there are 500 people in the population, there are 3 digits in the total, and there must be 3 digits in each name. Thus, the first case in the population is named 001.

that you wanted to conduct a survey of the opinions on "date rape" held by all students on a college campus. If you suspect that males and females might differ in their opinions, it would be desirable to first stratify the population according to gender and then draw separately from each stratum at random. Specifically, you would draw a random sample of males and separately draw a random sample of females. The same percentage should be drawn from each stratum. For example, if you want to sample 10% of the population and there are 1,600 males and 2,000 females, you would draw 160 males and 200 females. Notice that there are more females in the sample than males, which is appropriate because the females are more numerous in the population. It is important to notice that we are *not* stratifying in order to compare males with females; rather, we are attempting to obtain a sample of the entire college population that is representative in terms of gender.[3] With stratified random sampling, we have retained the benefits of randomization (i.e., the elimination of bias) and have gained the advantage of having appropriate proportions of males and females.[4] If your hunch was correct that males and females differ, you would have increased the precision of your results by stratifying.

Note that stratifying does not eliminate sampling errors. For example, when you drew the females at random, you may have, by chance, obtained females for your sample that are not representative of all females on the campus; the same, of course, holds true for men. However, you have eliminated sampling errors associated with gender because if you had used *simple random sampling,* you could have obtained a disproportionately large number of either males or females.

For large-scale studies, ***multistage random sampling*** may be used. In this technique, you might draw a sample of counties at random from all counties in the country, then draw voting precincts at random from all precincts in the counties selected, and finally draw individual voters at random from all precincts that were sampled. In multistage random sampling, you could introduce

---

[3]If your purpose were to compare males with females, then it would be acceptable to draw the same number of each and compare averages or percentages for the two samples.

[4]Of course, if your hunch that males and females differ in their opinion was wrong, the use of stratification would be of no benefit, but it would not introduce any additional errors beyond the sampling errors created at random.

stratification. For example, you could first stratify the counties into rural, sub-urban, and urban and then separately draw counties at random from these three types of counties—ensuring that all three types of counties are included.

A technique that is often useful is ***cluster sampling*** if it is done at random. To use cluster sampling, all members must belong to a cluster (i.e., an existing group). For example, all Boy Scouts belong to a troop, all students belong to a homeroom, etc. Unlike simple random sampling in which individuals are drawn, in cluster sampling *clusters* are drawn. To conduct a survey of Boy Scouts, for example, one could draw a random sample of troops, contact the leaders of the selected troops, and ask them to administer the questionnaires. The advantages are obvious; there are fewer people to contact (only the leaders) and the degree of cooperation is likely to be greater if a leader asks the Scouts to participate.[5] There is a disadvantage, however, which results from the fact that clusters often are homogeneous in some way. This disadvantage is illustrated best by example. Suppose that you drew 10 troops (i.e., clusters) at random and nine of them, by chance, were in major urban areas; Scouts in major urban areas may have different attitudes and skills than those in rural areas. Even though there might be about 20 scouts in each troop, yielding responses from about 200 Scouts, the number 200 is misleading because the sample size is 10 and *not* 200. If you had drawn 200 Boy Scouts by simple random sampling, it is very unlikely that you would get such a disproportionate number of Scouts from urban areas, and the sample size would be 200. Furthermore, had you used stratified random sampling and stratified on the basis of geographical area, you could have physically prevented such an error. Thus, when using cluster sampling, it is desirable to use a large number of clusters to overcome the disadvantage.

Terms to Review Before Attempting Worksheet 28

simple random sampling, table of random numbers, stratified random sampling, multistage random sampling, cluster sampling

---

[5]Keep in mind that for ethical reasons, it is highly desirable to obtain the informed consent of the subjects or guardians of minors before conducting most studies.

Worksheet 28: A Closer Look at Sampling

Riddle: How are minds like parachutes?

DIRECTIONS: To find the answer to the riddle, write the answer to each question in the space immediately below it. The word in parentheses in the solution section next to the answer to the first question is the first word in the answer to the riddle, the word beside the answer to the second question is the second word, and so on.

1. Suppose that the size of a population is 50 and that the members of the population are numbered from 01 to 50. If you draw a sample starting at the beginning of the fifth row in Table 2 near the end of this book, what is the number of the second case drawn?

2. Using the information in Question 1, what is the number of the third case drawn?

3. Suppose that the size of a population is 352 and that they are numbered from 001 to 352. If you draw a sample starting with the first row in Table 2 near the end of this book, what is the number of the second case drawn?

4. Suppose that you wish to use stratified random sampling to obtain a sample from a population in which there are 600 freshmen, 500 sophomores, 450 juniors, and 400 seniors. Should you draw the "same number" *or* the "same percentage" from each grade level?

Worksheet 28 (Continued)

5. Suppose that you randomly draw 10% of the Democrats, then separately randomly draw 10% of the Republicans, and finally randomly draw 10% of all other registered voters from a population. Are you using "simple random sampling," "stratified random sampling," *or* "cluster sampling"?

6. Is the purpose of stratified random sampling, as described in this section, to produce two or more samples that are to be compared?

7. If you used cluster sampling and drew at random 30 clusters with 10 members each, is the sample size 30, 10, or 300?

8. If you take a master list of all class sections being taught on a college campus and draw 50 sections at random, are you using "simple random sampling," "stratified random sampling," *or* "cluster sampling"?

Solution section:

> 498 (lonely)   no (they)   same number (feeling)   30 (are)   10 (others)
>
> cluster sampling (open)   stratified random sampling (when)   58 (jumping)
>
> 25 (they)   17 (will)   300 (prophet)   same percentage (function)   83 (run)
>
> multistage random sampling (he)   088 (only)   simple random sampling (it)

Worksheet 28 (Continued)

Write the answer to the riddle here, putting one word on each line: _____ _____ _____ _____ _____ _____ _____ _____

"Whenever I'm overwhelmed by statistics
homework, and I need a sunnier outlook,
I just turn up the brightness control."

Section 29: Introduction to Probability

As a result of sampling at random from a population (see Sections 27 and 28), we introduce random or chance errors into our data; these are called *sampling errors*. Much of the rest of this book concerns how to assess the possible role of random errors on our observations of a random sample from a population. We will use probabilities[1] to make decisions regarding this matter.

We all use probabilities informally each day to make decisions. When we are considering crossing a street, we judge the odds (i.e., probability) that we will make it safely to the other side before approaching traffic reaches us. When we hear in the morning that the probability of rain that day is 60%, we make decisions on what to wear and whether to take an umbrella with us. In a more formal manner, using probability theory, scientists make decisions after assessing the odds that random errors may have created the differences or relationships that they are examining.

Here are a few basic *principles of probability*:

1. Mutually exclusive events that are the result of chance are independent of each other.

Mutually exclusive means completely separate and having no bearing on each other, so the principle is almost self-defining. Yet, its implication sometimes escapes people. For example, consider a state lottery in which the numbers 1 through 46 may be drawn. What are the odds that a 39 will be the first number drawn on a given week? Obviously, 1 in 46 or 1/46, which is equivalent to .02 (obtained by dividing 1 by 46). The following week all 46 numbers are again in play (a mutually exclusive event). What are the odds that a 39 will be the first number drawn? It is still 1 in 46 or .02. Thus, knowledge of what was drawn one week gives a player no advantage in selecting numbers the next week because the events are independent of each other.

---

[1]Probability is defined as the number of times something is likely to occur out of the total number of possibilities.

2. Sum the probabilities of separate (mutually exclusive) events to determine the probability that any of them will occur.

In a lottery with 46 numbers, the odds that a 39 will be drawn are 1/46 or .02. The odds that a 24 will be drawn are also 1/46 when all numbers are in play. Thus, the odds that a 39 *or* 24 will be drawn are 1/46 + 1/46 = 2/46 = 1/23 or .04. This makes sense if you think about it. Clearly, the probability that *either* a 39 *or* 24 will be drawn is twice as great as the odds that just one of them will be drawn.

3. If something is certain to occur, the probability is 1.00.

In a lottery with 46 numbers, the probability that one of the 46 will be drawn is 1.00.

4. If something is certain *not* to occur, the probability is 0.00.

If a lottery is canceled and no numbers will be drawn, the probability that any of them will be drawn is 0.00.

5. Calculate the product of the separate probabilities to determine the probability of their successive occurrence.

To keep it simple for instructional purposes, assume that after a number is drawn in a lottery, it is placed back in play so that there are 46 numbers that may be drawn each time a draw is made. What is the probability that a 39 *and* a 24 will both be drawn in *a given lottery*? The fifth principle tells us that it is 1/46 × 1/46, which equals 1/2116 or .00047. This is a highly unlikely event. If you have to correctly select 6 numbers out of the 46 that will be drawn in a given lottery in order to win, you can determine your odds by multiplying 1/46 by itself six times. If you try this using a typical calculator (multiplying .02 by .02 by .02, etc.), you will find that your calculator will very quickly run out of decimal places for zeros; in other words, your calculator will run out of room to hold all the zero place holders and tell you that your odds are essentially zero. Try it and see what happens. Of course, there is some extremely small probability that you will win, but a typical calculator cannot show such a small probability.

In the next section, we will examine how probabilities relate to the normal curve.

Terms to Review Before Attempting Worksheet 29

**sampling errors, principles of probability,
mutually exclusive**

"I used a $3,000 computer, a $1,200 laser printer,
and a $300 data-processing program ——
and I still got a D on my statistics project!"

Worksheet 29: Introduction to Probability

> *Riddle*: What principle applies to both levying taxes and shearing sheep?

DIRECTIONS: To find the answer to the riddle, write the answer to each question in the space immediately below it. The word in parentheses in the solution section next to the answer to the first question is the first word in the answer to the riddle, the word beside the answer to the second question is the second word, and so on.

1. If something is certain to occur, what is the probability that it will occur expressed as a proportion (to four decimal places)?

2. If you are certain that something will not occur, what is its probability expressed as a proportion?

3. What is the probability that a 2 *and* an 8 will be drawn in a given lottery in which there are 20 numbers in play on each draw?

4. What is the probability that a 2 *and* an 8 will be drawn in a given lottery in which there are 10 numbers in play on each draw?

5. Assume that a deck of 52 cards is shuffled and a king of hearts is drawn. Then the 52 cards are shuffled again and another card is drawn. Is the probability that a king of hearts will be drawn the second time the same as the probability that it would be drawn the first time?

Worksheet 29 (Continued)

6. If you know that the number 2 was drawn in a lottery one week, does this help you in selecting numbers in an independent lottery the next week?

7. If a deck of 52 cards is shuffled, what is the probability, expressed as a proportion, that a jack of diamonds will be drawn?

8. If a deck of 52 cards is shuffled, what is the probability, expressed as a proportion, that any of the four aces will be drawn?

9. If the probability that Event A will occur is .005 and the probability that Event B will occur is .01, which event has a greater probability of occurring?

Solution section:

> Event A (government)   1/52 (painful)   Event B (skin)   1.0000 (you)
>
> 0.2500 (kill)   0.0000 (should)   0.0025 (stop)   0.0500 (Congress)
>
> 0.0100 (when)   yes (you)   0.0192 (to)   no (get)   100% (likely)
>
> 0.0769 (the)   0.0050 (broke)   probability (screaming)

Write the answer to the riddle here, putting one word on each line: _____ _____ _____ _____

_____ _____ _____ _____

Notes:

Section 30: Probability and the Normal Curve

In the empirical approach to knowledge, we make observations and, based on them, make decisions. Based on previous observations, we can establish probabilities regarding the occurrence of specific events in the future. Weather forecasting is based on this approach. Events such as high and low pressure systems are observed, and predictions are made based on previous observations of their effects on the weather.

Fortunately, for many problems, empirical probabilities are easy to determine because many distributions are normal.[1] Suppose, for example, that we conducted a large national survey to determine knowledge of basic math skills; each subject was administered a basic math test. If we found that the distribution was normal and that the mean was 50.00 and the standard deviation was 7.00, we could use this information to establish probabilities. To do so, we would need to use z-scores. You may recall that the formula for them is:[2]

$$z = \frac{X - M}{S}$$

In this example, what is the probability of drawing an individual at random from the population who has a score of 64 or higher? To answer the question, first calculate the corresponding z-score:

$$z = \frac{64 - 50.00}{7.00} = \frac{14.00}{7.00} = 2.00$$

Then look up the z-score in Table 1 near the end of this book. There we find that the percent of cases in the smaller part (Column 4) is 2.28%. Divide this by 100, and we obtain a probability of .0228, which indicates that there is only slightly more than 2 chances in 100 of drawing such a person from the population. This is referred to as a ***one-tailed probability*** because we asked the

---

[1]The normal curve was first introduced in Section 9 and explored in Sections 13 through 16.
[2]See Section 16 to review z-scores.

question about only the upper tail of the normal distribution—the right-hand tail of the distribution in Figure 30.1.

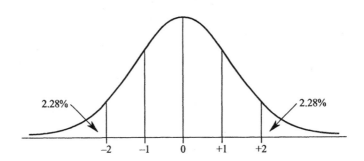

Figure 30.1. Normal distribution with selected *z*-scores.

Suppose, instead, we asked this question: What is the probability of drawing at random an individual with a *z*-score as high as 2.00 (or higher) *or* as low as –2.00 (or lower)? Obviously, the odds of doing so are double those of drawing just one of these. Thus, the odds are 2 × .0228 = .0456, which is just a little more than 4 in 100. This is called a ***two-tailed probability*** because we are asking about the odds of drawing an individual at either tail of the normal distribution in Figure 30.1. The importance of distinguishing between one-tailed and two-tailed probabilities will become clear in later sections.

The probabilities for both events described above are ***unlikely events***. In most sciences, conventional wisdom indicates that any event that has a probability of occurrence of .05 or less is usually classified as unlikely to occur at random. You will notice in Table 1, under *Values of Special Interest*, that a *z*-score of 1.96 has only a 2.5% chance of occurrence as a one-tailed probability (see Column 4); as a two-tailed probability, it has a 5.0% chance of occurrence (2 × 2.5% = 5.0%). Thus, an event with a *z*-score of 1.96 or greater or –1.96 or less (such as 1.97 or –1.97) is classified as an unlikely event.

Table 1 also indicates that a *z*-score of 2.58 or higher has only a .49% (or almost 1/2 of 1%) chance of occurrence. The corresponding two-tailed probability is .98% or almost 1% for *z*-scores of 2.58 or –2.58. Some scientists *only* classify an event as unlikely if its likelihood is 1% or less.

There is no rule of nature that says at what point an event should be classified as unlikely. However, the 5% and 1% guidelines have evolved over time as the two most widely used. The only universally accepted rule is that a researcher must decide *in advance* of examining the data what guideline will be followed for declaring an event to be unlikely. Theoretically, any percentage may be specified, but, to be accepted in most scientific circles, 5% or lower is generally used.

Identifying unlikely events is the basis of many of the tests of statistical significance presented in later sections.

Terms to Review Before Attempting Worksheet 30

one-tailed probability, two-tailed probability, unlikely events

**"I have a photographic memory—
I just seem to be out of film today."**

Worksheet 30: Probability and the Normal Curve

Riddle: What did *George Washington* do when he was asked for his ID?

DIRECTIONS: To find the answer to the riddle, write the answer to each question in the space immediately below it. The word in parentheses in the solution section next to the answer to the first question is the first word in the answer to the riddle, the word beside the answer to the second question is the second word, and so on.

1. What is the one-tailed probability of drawing a subject with a z-score of 1.35 or higher at random from a normal distribution? (Hint: Remember to divide the percentage you obtain from Table 1 by 100.)

2. What is the one-tailed probability of drawing a subject with a z-score of -1.70 or lower at random from a normal distribution?

3. What is the two-tailed probability of drawing a subject with a z-score as extreme as 1.80 or -1.80 at random from a normal distribution?

4. For a normal distribution with a mean of 100.00 and a standard deviation of 16.00, what is the one-tailed probability of drawing a subject with a score of 124 or greater at random from a normal distribution?

Worksheet 30 (Continued)

5. For a normal distribution with a mean of 40.00 and a standard deviation of 8.00, what is the one-tailed probability of drawing a subject with a score of 30 or less at random from a normal distribution?

6. Using the information in Question 5, what is the probability of drawing a subject with a score as extreme as 30 or 50 at random from a normal distribution?

7. Using the 5% guideline, should the answer to Question 2 be classified as an unlikely event?

8. Using the 1% guideline, should the answer to Question 5 be classified as an unlikely event?

Solution section:

.4554 (Congress)   no (it)   .0359 (public)   yes (showed)   .0718 (out)

.2112 (and)   .1336 (fooling)   .1056 (quarter)   .1770 (deficit)

.0668 (a)   .0885 (he)   .9554 (politicians)   .0446 (whipped)   .4115 (saying)

Write the answer to the riddle here, putting one word on each line: _____ _____ _____ _____
_____ _____ _____ _____

Notes:

Section 31: Standard Error of the Mean

Suppose that there is a large population with a mean of 100.00 and a standard deviation of 16.00 on a standardized test. Further, suppose that we do not have this information but wish to estimate the mean and standard deviation by testing a sample. When we draw a random sample and administer the test to just the sample, will we correctly estimate the population mean as 100.00? The answer is probably not. Remember that random sampling introduces random or chance errors—known as sampling errors.

At first, this situation may seem rather hopeless, but we have the advantage of using an unbiased, random sample—meaning that no factors are systematically pushing our estimate in the wrong direction. In such a situation, using a large sample increases the likelihood that we are correct or, at least, not likely to make a large error.[1]

We also have the advantage of the *central limit theorem*. To understand this theorem, you must first understand the sampling distribution of means. Suppose that we drew not just one sample but many samples at random. That is, we drew a sample of 60, tested the subjects, and computed the mean; then drew another sample of 60, tested the subjects, and computed the mean; then drew another sample of 60, tested the subjects, and computed the mean; etc. We would then have a very large number of means—known as the *sampling distribution of means.* *The central limit theorem says that the distribution of these means is normal in shape.* The normal shape will emerge even if the underlying distribution is skewed, provided that the sample size is reasonably large (about 60 or more). The mean of an indefinitely large sampling distribution of means will equal the population mean. The standard deviation of the sampling distribution is known as the *standard error of the mean*. Keep in mind that the means vary from each other only because of chance errors created by random sampling. That is, we are drawing random samples from the same population and administering the same test over and over, so all of the means should have the same value except for the effects of random errors.

---

[1]Review Sections 27 and 28 on sampling.

Therefore, there is variation among the means only because of sampling errors. For this reason, the *standard deviation of the sampling distribution* is known as the *standard error of the mean.*

In practice, we usually draw a single sample, test it, and calculate its mean and standard deviation.[2] Therefore, we are not certain of the value of the population mean nor do we know the value of the standard error of the mean that we would obtain if we had sampled repeatedly. Fortunately, we do know two very useful things:

➡ The larger the sample, the smaller the standard error of the mean.

➡ The less the variability in the population, the smaller the standard error of the mean. For example, consider a population in which there is no variability—that is, in which all subjects are identical. In this case, the standard error of the mean (i.e., the standard deviation of the sampling distributions of means) equals 0.00 (i.e., all the means will be identical and their standard deviation will be zero). In practice, we cannot be certain how much variability there is in a population from which we have only sampled. However, we can use the standard deviation of the sample that we have drawn as an estimate of the amount of variability in the population; for example, if we observed a very small standard deviation for a random sample, it would be reasonable to guess that the population has relatively little variation.

Given these two facts and some statistical theory that is not covered here, statisticians have developed this formula for estimating the standard error of

---

[2]Use the second formula in Appendix A when estimating the standard deviation of a population from a sample.

the mean based only on the information we have about a given random sample from a population:

$$SE_m = \frac{s}{\sqrt{n}}$$

Let's apply the formula in three examples to see how it works:

Example 1:

For a randomly selected sample, $m = 75.00$, $s = 16.00$, and $n = 64$. If we divide 16.00 by the square root of 64 (i.e., 8), we estimate that the standard error of the mean equals 2.00. This is an estimate of a ***margin of error*** that we should keep in mind when interpreting the sample mean of 75.00.

Keep in mind, too, that the *standard error of the mean* is an estimate of the *standard deviation of the sampling distribution of the means*, which is normal in shape when the sample size is relatively large. You may recall from your study of the standard deviation that about 68% of the cases lie within one standard deviation unit of the mean. Thus, we would expect about 68% of all sample means to lie within 2.00 points of the true (or population) mean. If we use the mean of 75.00, which was actually obtained, as an estimate of the population mean based on a random sample, we could estimate that odds are 68 out of 100 that the population mean lies between 73.00 (75.00 − 2.00 = 73.00) and 77.00 (75.00 + 2.00 = 77.00). The values of 73.00 and 77.00 are known as the ***limits of the 68% confidence interval for the mean***. That is, we have about 68% confidence that the true mean lies between 73.00 and 77.00.

Example 2:

For a randomly selected sample, $m = 75.00$, $s = 16.00$, and $n = 128$. If we divide 16.00 by the square root of 128 (i.e., 11.314), we estimate that the standard error of the mean equals 1.41. Notice that this is substantially smaller than the standard error we obtained in Example 1. This is because the sample size is twice that in Example 1. However, also notice that the standard error has not been cut in half. This is because we are dividing by the *square root* of *n*.[3]

The limits of the 68% confidence interval for Example 2 are 73.59 (75.00 − 1.41 = 73.59) and 76.41 (75 + 1.41 = 76.41). The larger sample size in Example 2 has given us a smaller confidence interval than in Example 1.

Example 3:

For a randomly selected sample, $m = 75.00$, $s = 5.00$, and $n = 128$. If we divide 5.00 by the square root of 128 (i.e., 11.314), we estimate that the standard error of the mean equals 0.44. The limits of the 68% confidence interval are 74.56 and 75.44. The smaller standard deviation in Example 3 has given us a smaller confidence interval than in Example 2.

It should be obvious that a small confidence interval is desirable because it indicates that the sample mean is probably close to the true mean. Of the two variables that affect the size of the standard error of the mean—the sample size and the variability of the sample—we often have direct control of the sample size. By using reasonably large samples, we can minimize the standard error of the mean.

Of course, 68% confidence is far short of certainty. The next section covers how to build 95% and 99% confidence intervals—intervals within which we can have much greater confidence that the population mean lies.[4]

It is important to keep in mind that the confidence limits are only valid if we analyze the results obtained with unbiased (random) sampling. Each bias has its own unique and usually unknown effects on the results, and there are no generalizable techniques for estimating the amount of error created by them.

---

[3]You may recall from Section 27 that increasing the sample size produces diminishing returns, which is clearly illustrated here.

[4]Appendix F presents the formulas for calculating the standard error of a median and the standard error of a percentage.

Terms to Review Before Attempting Worksheet 31

**central limit theorem, sampling distribution of means,
standard error of the mean, margin of error,
limits of the 68% confidence interval for the mean**

"Statistics show that to prevent a heart attack, you should
take one aspirin every day. Take it out for a jog, then
take it to the gym, then take it for a bike ride...."

Worksheet 31: Standard Error of the Mean

Riddle: According to Frank Lloyd Wright, how are the truth and the facts related?

DIRECTIONS: To find the answer to the riddle, write the answer to each question in the space immediately below it. The word in parentheses in the solution section next to the answer to the first question is the first word in the answer to the riddle, the word beside the answer to the second question is the second word, and so on.

1. What is the name of the theorem that says a large sampling distribution of means will be normal in shape?

2. "Other things being equal, the larger the sample size, the larger the standard error of the mean." Is this statement true or false?

3. "Other things being equal, the greater the variability in a population, the greater the amount of error that can be expected when we sample." Is this statement true or false?

4. If we double the size of a sample, can we expect to have half the amount of error due to random sampling?

5. For a randomly selected sample, $m = 50.00$, $s = 10.00$, and $n = 64$. What is the value of the standard error of the mean?

Worksheet 31 (Continued)

6. For a randomly selected sample, $m = 50.00$, $s = 10.00$, and $n = 144$. What is the value of the standard error of the mean?

7. For a randomly selected sample, $m = 100.00$, $s = 16.00$, and $n = 100$. What are the limits of the 68% confidence interval for the mean?

8. For a randomly selected sample, the mean equals 90.00 and the standard error of the mean equals 5.22. What are the limits of the 68% confidence interval for the mean?

Solution section:

84.78–95.22 (facts) standard error theorem (justice) 98.40–101.60 (the)		
yes (Bible) 1.25 (important) 5.00 (lawyer) 0.15 (court)		
no (more) true (is) false (truth) central limit theorem (the) 0.83 (than)		
4.17 (being) 6.25 (building) 0.16 (commandment) 84.00–116.00 (seeking)		

Write the answer to the riddle here, putting one word on each line: _____ _____ _____ _____
_____ _____ _____ _____

Notes:

Section 32: Confidence Interval for the Mean

In the previous section, you learned how to compute the standard error of the mean (SE_M) and how to use it to calculate the limits of the 68% confidence interval for a mean. Often, we want to calculate limits in which we can have more than 68% confidence. This section shows you how to do this, provided you have a reasonably large sample size (about 60 or more).[1]

To determine the limits of the **95% confidence interval** for a mean, first multiply the standard error of the mean by the constant 1.96 and then add it to and subtract it from the mean.[2]

Example 1:

If m = 40.00 and SE_m = 1.50, then the limits of the 95% confidence interval for the mean are obtained as follows:

FIRST: (1.50)(1.96) = 2.94

SECOND: 40.00 – 2.94 = 37.06

THIRD: 40.00 + 2.94 = 42.94

THUS, the limits of the 95% confidence interval are 37.06 and 42.94. The **lower limit** is 37.06 and the **upper limit** is 42.94. We can state that we have **95% confidence** that the true (i.e., population) mean lies between 37.06 and 42.94.

The process followed in Example 1 to obtain the 95% confidence interval is expressed by this formula:

$$CI_{95} = m \pm (1.96)(SE_m)$$

---

[1] With 60 or more, your answers will be very close to the precise answer. Refer to Appendix G to learn how to build intervals when your sample size is less than 60.

[2] You may recall the value of 1.96 from the table of the normal curve (see Table 1 near the end of this book). If you go out 1.96 standard deviation units from the mean in both directions, you capture 95% of the cases in a normal distribution. You may also recall that the sampling distribution of the mean is normal when the sample size is reasonably large.

Note that the limits of the 68% confidence interval for Example 1 are 38.50 (40.00 − 1.50) and 41.50 (40.00 + 1.50). If we are willing to settle for less confidence (68% rather than 95%), we obtain a smaller interval. Put another way, the greater the confidence desired, the larger the interval. This makes sense. To obtain a greater degree of confidence for a given set of data, we have to allow for a larger interval as our estimate of where the true mean lies. By having a larger interval, we include more possibilities and can be more confident that one of these possibilities is the true mean.

To determine the limits of the **99% confidence interval** for a mean, first multiply the standard error of the mean by the constant 2.58 and then add it to and subtract it from the mean.[3]

Example 2:

If m = 40.00 and SE_m = 1.50, then the limits of the 99% confidence interval for the mean are obtained as follows:

FIRST: (1.50)(2.58) = 3.87

SECOND: 40.00 − 3.87 = 36.13

THIRD: 40.00 + 3.87 = 43.87

THUS, the limits of the 99% confidence interval are 36.13 and 43.87. The *lower limit* is 36.13 and the *upper limit* is 43.87. We can state that we have *99% confidence* that the true (i.e., population) mean lies between 36.13 and 43.87.

The process followed in Example 2 to obtain the 99% confidence interval is expressed by this formula:

$$CI_{99} = m \pm (2.58)(SE_m)$$

Notice that the interval for 99% confidence is larger than the interval for 95% confidence, illustrating once again that for a higher degree of confidence, we have to allow for a greater range of possibilities.

[3]You may recall from Table 1 near the end of this book that if you go out 2.58 standard deviation units on both sides of the mean in a normal distribution, you capture 99% of the cases.

We have a very high degree of confidence in the result of Example 2. We can obtain such a high degree of confidence within a rather small range of possibilities by using a reasonably large sample. Remember that the larger the sample, the smaller the standard error of the mean. With a smaller standard error, the resulting confidence interval will be smaller.

At first, some of you may wonder why we do not determine the interval in which we have *100% confidence*. Brief reflection on the problem reveals that this is impossible if we are to have a useful answer. Because random sampling introduces errors whose effects we can only estimate, the only way to have 100% confidence is to build an interval that includes all of the possibilities. Thus, if we are giving a test with possible scores from zero to 100, we can have 100% confidence that the population mean is either zero, 100, or somewhere in between. We know this before we test a single subject, and, thus, this result is not useful.

Appendix G describes how to build confidence intervals when small samples (e.g., samples of 60 or less) are used.

Terms to Review Before Attempting Worksheet 32

**95% confidence interval, 99% confidence interval,
lower limit, upper limit**

Worksheet 32: Confidence Interval for the Mean

Riddle: According to Seneca, why have many people failed to attain wisdom?

DIRECTIONS: To find the answer to the riddle, write the answer to each question in the space immediately below it. The word in parentheses in the solution section next to the answer to the first question is the first word in the answer to the riddle, the word beside the answer to the second question is the second word, and so on.

1. For a given set of data, will the 95% or the 99% confidence interval be larger?

2. For a given set of data, will the 68% or the 95% confidence interval be smaller?

3. "Other things being equal, the larger the sample size, the smaller the confidence interval for the mean." Is this statement true or false?

4. If $m = 50.00$ and $SE_m = 3.00$, what are the limits of the 95% confidence interval for the mean?

5. If $m = 50.00$ and $SE_m = 3.00$, what are the limits of the 99% confidence interval for the mean?

6. Is the interval in the answer to "Question 4" *or* "Question 5" larger?

Worksheet 32 (Continued)

7. If $m = 40.00$, $s = 12.00$, and $n = 144$, what are the limits of the 95% confidence interval for the mean? (Review Section 31 before answering this question.)

8. For the data in Question 7, what are the limits of the 99% confidence interval for the mean?

9. Is the interval in the answer to Question 7 or Question 8 smaller?

Solution section:

95% (youngster)	false (wisdom)	Question 4 (silly)	99% (because)
true (foolishly)	68% (they)	Question 5 (they)	42.26–57.74 (that)
44.12–55.88 (assumed)	28.00–52.00 (birthday)	Question 7 (it)	
Question 8 (helpless)	37.42–42.58 (possessed)	38.04–41.96 (already)	

Write the answer to the riddle here, putting one word on each line: _____ _____ _____ _____ _____ _____ _____ _____ _____

Notes:

Section 33: Introduction to the Null Hypothesis

   Suppose that we draw a random sample of first-grade girls and a random sample of first-grade boys from a large school district in order to estimate the average reading achievement of both groups on a standardized test. Let us suppose we obtain these means:

Girls	Boys
$m = 50.00$	$m = 46.00$

This result suggests that girls, on the average, have higher achievement in reading. But do they? Remember that we have tested only a random sample of the boys and girls. It is possible that the difference that we obtained is due only to the errors created by random sampling, which are known as **sampling errors**. In other words, it is possible that the population mean for boys and the population mean for girls are identical, and we found a difference only because of the effects of random sampling. This possibility is known as the **null hypothesis**. For the difference between two means, it says that:

   The true difference between the means (in the population) is zero.

This statement can also be expressed with symbols, as follows:[1]

   $H_0: \mu_1 - \mu_2 = 0$
         Where:

               H_0 is the symbol for the null hypothesis.
               μ_1 is the symbol for the *population* mean for one group.
               μ_2 is the symbol for the *population* mean for the other group.

Another way to state the null hypothesis is:

   Another expression of the null hypothesis, which underlies a one-tailed test, is beyond the scope of this section but is discussed in Section 35.

There is no true difference between the means.

The null hypothesis may also be stated in the positive as follows:

The observed difference between the means was created by sampling error.

Most investigators are searching for differences among people and for ex-planations for the differences that they find. Therefore, most do not undertake their studies in the hope of confirming the null hypothesis. Yet, once they have sampled at random, they are *stuck* with the null hypothesis as a possible expla-nation for any observed differences. They may also have their own personal hypothesis (i.e., their **research hypothesis**) that is not consistent with the null hypothesis. One possibility is that they believe that the average reading achievement of girls is higher than that of boys. This is known as a **directional hypothesis** because it states that one particular group's average is higher than the other. Expressed as symbols, this hypothesis is:

$H_1: = \mu_1 > \mu_2$
    Where:

        H_1 is the symbol for an **alternative hypothesis** (i.e., an al-ternative to the null hypothesis).
        μ_1 is the symbol for the *population* mean for the group hy-pothesized to have a higher mean (in this case, the girls).
        μ_2 is the symbol for the *population* mean for the other group (in this case, the boys).

Another investigator may hold a **nondirectional hypothesis** as his or her *re-search hypothesis*. That is, he or she believes that there is a difference between boys' and girls' reading achievement, but that there is insufficient information to hypothesize as to which group is higher. In other words, the investigator is hypothesizing that there is a difference—that the two groups are not equal

—but is not willing to speculate in advance on the direction of the difference. This is how to state a nondirectional research hypothesis in symbols:

$H_1: = \mu_1 \neq \mu_2$

Where:

H_1 is the symbol for an *alternative hypothesis* (i.e., an alternative to the null hypothesis).

μ_1 is the symbol for the *population* mean for one group.
μ_2 is the symbol for the *population* mean for the other group.

Because it is easy to get lost the first time, let us reconsider the possibilities. An investigator may conduct research in which two means are compared because he or she hypothesizes one of three things:

1. There is no difference. (This assertion is consistent with the null hypothesis, but is not frequently held.)
2. One specific group is higher than the other. (This is the most frequently held. It is a *directional* hypothesis.)
3. There is a difference between the two groups in an unspecified direction. (This assertion is not frequently held. It is a *nondirectional* hypothesis.)

Whichever hypothesis an investigator believes is true at the onset of the study is his or her *research hypothesis*. As you can see above, the most frequently held is number 2, a *directional research hypothesis*. Let us suppose that the two means we considered at the beginning of this section (i.e., $m = 50.00$ for girls and $m = 46.00$ for boys) were obtained by an investigator who started with the directional research hypothesis that girls, on the average, achieve more in reading than boys. Clearly, the observed means support the research hypothesis, but is the investigator finished? Obviously not—because he or she has two possible explanations for the observed difference:

1. Girls have higher achievement in reading than boys. (This is the research hypothesis.)
2. The observed difference in the sample is the result of the effects of random sampling; therefore, there is no true difference. (This is the null hypothesis.)

Notice that the researcher is at least temporarily *stuck* with the null hypothesis. Like it or not, because only a random sample was studied, he or she may be observing a difference that is the result of sampling errors. Thus, the null hypothesis is a possible explanation for the difference. If the investigator stops at this point, he or she has two explanations for a single difference of four points. This is hardly a definitive result. This investigator obviously should try to rule out the null hypothesis, leaving only the original research hypothesis. With the aid of inferential statistics, the investigator may test the null hypothesis. As a result of the inferential test, he or she may be able to rule it out.

The rest of this book deals with the null hypothesis and tests of it. In the next section, we will examine the logic of the null hypothesis in more detail.

Terms to Review Before Attempting Worksheet 33

**sampling errors, null hypothesis, research hypothesis,
directional hypothesis, alternative hypothesis, nondirectional hypothesis**

Worksheet 33: Introduction to the Null Hypothesis

Riddle: According to Mark Twain, why is it better to keep your mouth shut and appear stupid?

DIRECTIONS: To find the answer to the riddle, write the answer to each question in the space immediately below it. The word in parentheses in the solution section next to the answer to the first question is the first word in the answer to the riddle, the word beside the answer to the second question is the second word, and so on.

1. An investigator has studied *all* girls and boys in a population. She has found a difference between the mean for boys and the mean for girls. Is the null hypothesis a viable hypothesis?

2. What is the name of the hypothesis that says: "The true difference between the means equals zero"?

3. What is the name of the type of alternative hypothesis that states that Group X, on the average, is different from Group Y in an unspecified direction?

4. What is the name of the type of alternative hypothesis that is more frequently held by investigators?

5. What is the symbol for the null hypothesis?

6. What is the symbol for an alternative hypothesis?

Worksheet 33 (Continued)

7. For what does the symbol μ stand?

8. Is the assertion that "there is no difference" an *infrequently* held research hypothesis?

9. What term is defined as "errors created by random sampling"?

Solution section:

> alternative hypothesis (cows) H_n (marriage) M (tractor)
>
> sampling errors (doubt)   yes (all)   no (if)   null hypothesis (you)
>
> nondirectional hypothesis (open)   population mean (remove)
>
> H_1 (might)   directional hypothesis (it) H_0 (you)

**Write the answer to the riddle here, putting one word on each line: _____ _____ _____ _____,
_____ _____ _____ _____ _____**

Section 34: Decisions About the Null Hypothesis

In the previous section, you learned that the null hypothesis states that there is no true difference between two means—that in the population, the difference is zero.[1] In other words, a difference between means was obtained only because of sampling errors created by random sampling. As you will see later in this book, the same null hypothesis applies if you are comparing a variety of other statistics such as two frequencies (e.g., the frequency who said "yes" vs. the frequency who said "no").

The sections that follow deal with how to test the null hypothesis. The tests vary according to which statistics are being compared. But all tests have the same underlying logic. Let us examine the logic before proceeding with the mechanics of specific tests.

An inferential test of a null hypothesis yields, as its final result, a ***probability*** *that the null hypothesis is true*. The symbol for the probability is a lower-case ***p***. Thus, if we find that the probability that the null hypothesis is true in a given study is less than 5 in 100, this result would be expressed as $p < .05$. How should this be interpreted? What does it tell us about the null hypothesis? Quite simply, it tells us that it is *unlikely* that the null hypothesis is true. If it is unlikely to be true, what should we conclude about it? That it is probably not true.

It is important to understand the point of the previous paragraph, so let us consider an analogy. Suppose the weather forecaster reports that the probability of rain tomorrow is less than 5 in 100. What should we conclude? First, we know that there is some chance of rain—but it is very small. Because of its low probability, most people would conclude that it probably will not rain and not make any special preparations for it. By not making any special preparations, they are acting as though it will not rain; for all practical purposes, they have rejected the hypothesis that it will rain tomorrow.

There is always some probability that the null hypothesis is true—so if we wait for certainty, we will never be able to make a decision. Thus, statisticians

---

[1]For one-tailed tests described in Section 36, the null hypothesis is expressed in a different form.

and applied researchers have settled on the .05 level as the level at which it is appropriate to reject the null hypothesis.[2] When an alpha of .05 is used, we are, in effect, willing to be wrong 5 times in 100 in rejecting the null hypothesis. Consider the rain analogy. If we make no special preparations for rain 100 days for which the probability of rain is .05, it will probably rain 5 of those 100 days. Thus, in rejecting the null hypothesis, we are taking a calculated risk that we might be wrong. This type of error is known as a ***Type I error***—the error of rejecting the null hypothesis when it is correct. On those 5 days in 100 when you are caught in the rain without your rain gear, you will get wet because you made Type I errors.

In review, when the probability is low that the null hypothesis is correct, we reject the null hypothesis. A synonym for rejecting the null hypothesis is declaring a result to be ***statistically significant***. In academic journals, you will find statements such as this: *The difference between the means is statistically significant*. This means that the authors have rejected the null hypothesis.

In journals, you will frequently find p values of less than .05 reported. The most common are $p < .01$ (less than 1 in 100) and $p < .001$ (less than 1 in 1,000). When a result is statistically significant at these levels, investigators can be more confident that they are making the right decision in rejecting the null hypothesis than they could be by using the .05 level. Clearly, if there is only 1 chance in 1,000 that something is true, it is less likely that it is true than if there are 5 chances in 100 that it is true. For this reason, the .01 level is a *higher* level of significance than the .05 level, and the .001 level is a *higher* level of significance than the .01 level. Let us review:

.06+ level: *not* significant; do *not* reject the null hypothesis.

.05 level: significant; reject the null hypothesis.

.01 level: more significant; reject the null hypothesis with more confidence than at the .05 level.

.001 level: highly significant; reject the null hypothesis with even more confidence than at the .01 or .05 levels.

---

[2]The probability at which we are willing to reject the null hypothesis is known as the ***alpha*** level.

So what probability level should be used? Remember that most investigators are looking for significant differences (or relationships). Thus, they are most likely to use the .05 level because this is the easiest to achieve.[3]

Should you decide to use some level other than .05, you should decide that in advance of examining the data. Keep in mind, though, that when you require a lower probability before rejecting the null hypothesis (e.g., .01 instead of .05), you are increasing the likelihood that you will make a *Type II error*.[4] A Type II error is the error of failing to reject the null hypothesis when it is false. This type of error can have serious consequences. Suppose a drug company has developed a new drug for a serious disease. Suppose that, in reality, the new drug is effective. If, however, the null hypothesis is not rejected because the drug company selected a level of significance that is too high, the results of the study will have to be described as insignificant, and the drug may not receive government approval.

In review, there are two types of errors that can be made when making a decision about the null hypothesis:

> *Type I error*: reject the null hypothesis when, in reality, it is true.
> *Type II error*: fail to reject the null hypothesis when, in reality, it is false.

At first, this might seem frustrating—we have done our best, but we are still faced with the possibility of errors. While this is true, we will be making informed decisions, in the face of uncertainty, by using probabilities to our advantage. Either decision we make about the null hypothesis (reject or fail to reject) may be wrong, but by using inferential statistics to make the decisions, we can report to our readers the probability that we have made a *Type I error* (indicated by the p value we report). By reporting the probability level that we used,

---

[3]However, if they find that their result is significant at the .01 or .001 levels, they will report it at these levels for the readers' information. However, if it had only reached the .05 level, they still would have reported it as significant in many cases. The next section provides more information.
[4]The probability of this type of error is known as *beta*.

readers will be informed of the likelihood that we were incorrect when we decided to reject the null hypothesis.

In the next section, we will consider how to conduct a significance test for the difference between two means. The results of our computations will lead to a value of p, and when it is equal to or less than .05 (or some other level selected in advance), we will reject the null hypothesis.

Terms to Review Before Attempting Worksheet 34

probability, p, alpha, Type I error, statistically significant, Type II error, beta

Worksheet 34: Decisions About the Null Hypothesis

> **Riddle:** According to Robert Quillen, what is a good reducing exercise?

DIRECTIONS: To find the answer to the riddle, write the answer to each question in the space immediately below it. The word in parentheses in the solution section next to the answer to the first question is the first word in the answer to the riddle, the word beside the answer to the second question is the second word, and so on.

1. What is the symbol for probability?

2. What is the name of the error of rejecting the null hypothesis when it is true?

3. If a difference is declared to be statistically significant, is the null hypothesis rejected?

4. Is the .05 level or the .01 level a higher level of significance?

5. "When $p = .10$, the null hypothesis is usually rejected." Is this statement true or false?

6. "When $p = .05$, the null hypothesis is usually rejected." Is this statement true or false?

Worksheet 34 (Continued)

7. Is it possible to reject the null hypothesis with 100% certainty?

8. What is the name of the error of failing to reject the null hypothesis when it is, in reality, false?

9. Is a difference usually regarded as "statistically significant" *or* "statistically insignificant" when $p < .01$?

Solution section:

statistically significant (back)	Type II error (pushing)	P (jumping)
.05 (running) no (and) p (placing)		Type I error (both)
statistically insignificant (eating) true (table)		false (the)
.01 (against) yes (hands) null hypothesis (calories)		.10 (feast)

Write the answer to the riddle here, putting one word on each line: _____ _____ _____ _____
_____ _____ _____ _____ _____

Section 35: *z* Test for One Sample

The national mean for the population who takes the College Board's *SAT-Verbal* is 500.00 and the standard deviation is 100.00. Using the symbols for the population mean and standard deviation, we can say that:

$$\mu = 500.00 \text{ and } \sigma = 100.00$$

(Note that μ is the symbol for the population mean and σ is the symbol for the population standard deviation.)

Suppose that researchers for the Department of Education of the State of Disrepair (the 51st state admitted to the Union) suspected that their students, on the average, perform more poorly than the national population. (This is their *research hypothesis*.) Suppose they drew a random sample of 200 students who took the test and found that:

$$m = 485.00 \text{ and } s = 101.00$$

At first glance, the data seem to support their *research hypothesis*; on the average, the sample of students from the state is 15 points below the national population. However, the *null hypothesis* also offers an explanation for the 15-point difference. It states that the difference was created by sampling errors due to the random sampling—that, in fact, the *true* difference is zero. (It asserts that this true difference of zero would have been obtained if the entire population, instead of just a random sample, had been studied.) Thus, we have one difference of 15 points for which we have two explanations:

        1. the *research hypothesis* and

        2. the *null hypothesis*.

To determine whether the *null hypothesis* is viable, we can test it with a *z test*.

You probably recall that a *z*-score for an individual is computed using this formula:[1]

$$z = \frac{X - m}{s}$$

---

[1] See Section 16 to review *z*-scores.

207

It indicates how many standard deviations a person is from the mean of his or her group; *z*-scores of greater than 1.96 and less than −1.96 occur less than 5% of the time in a normal distribution. Thus, we can say that the probability of drawing a person on a single random draw who has a *z*-score this extreme is an unlikely event.[2] We can use the same logic (but a modified formula) to determine whether the sample mean of 485.00 is an unlikely event; that is, it is unlikely to be obtained by random sampling from a population with a mean of 500.00. The formula we use for the *z* test is:

$$z = \frac{m - \mu}{SE_m}$$

The denominator of the formula should look familiar. It is the symbol for the standard error of the mean, which we examined in Section 31. Because we know the standard deviation of the population, we use it to calculate SE_m instead of the standard deviation of the sample because the standard deviation of the population is a value that we know is free of sampling errors while the standard deviation of the sample is subject to such errors. Thus, we first calculate:

$$SE_m = \frac{\sigma}{\sqrt{n}} = \frac{100.00}{\sqrt{200}} = \frac{100.00}{14.142} = 7.071$$

Substituting, we obtain:

$$z = \frac{485.00 - 500.00}{7.071} = \frac{-15.00}{7.071} = -2.121$$

Now that we have the value of *z* for this *z* test, we evaluate it to determine if it is an unlikely event. If we determine that a mean of 485.00 is unlikely to be obtained by random sampling from a population with a mean of 500.00, we

---

[2]See Section 30 to review this concept.

will *reject the null hypothesis* and declare the difference to be *statistically significant*.

Remember that the null hypothesis says that the mean difference of 15 points is merely due to random sampling errors. If we determine that this is unlikely, we will reject the hypothesis. In statistics, as in everyday life, if something is unlikely to be true, we reject it and act as though it is false.

To evaluate our *z* of −2.121, we will first use the constants 1.96 and −1.96. The table of the normal curve (see Section 30 and Table 1 near the end of this book) tells us that the probability of obtaining a *z* this extreme is .05 or 5 in 100. Because we obtained a *z* of −2.121, the odds of obtaining our particular result are *less than* 5 in 100. This result may be reported in one of two ways. Note that they both have the same meaning and implications:

1. The null hypothesis has been rejected at the .05 level.
2. The difference is statistically significant at the .05 level.

We can also evaluate our value of *z* using the constants of 2.58 and −2.58. As you learned in Section 30, the odds of obtaining a *z* this extreme are .01 or 1 in 100. Because we obtained a *z* of −2.121, our result is *not* sufficiently extreme to classify this as an unlikely event at the .01 level. Thus, using this level, we report that:

1. The null hypothesis has *not* been rejected at the .01 level. (Another way of saying this is: We have failed to reject the null hypothesis at the .01 level.)
2. The difference is *not* significant at the .01 level. (Another way of saying this is: The difference is insignificant at the .01 level.)

Notice that the two results we have just examined are not contradictions. In practice, before examining the data, you should select a level (usually .05 or .01) that you will use in your significance test. Had you chosen the .05 level, you would report to your audience that the difference is significant at that level. Had you initially chosen the .01 level, you would report that it is *not* significant at that level. Initially, you should choose only one level.

It is important to note that the decision to *not reject* the null hypothesis is *not* equivalent to *accepting* the null hypothesis. Recall at the beginning of this section that we had two hypotheses that might explain the difference—the null hypothesis and the research hypothesis. If we fail to reject the null hypothesis, we are still left with two hypotheses—the null hypothesis, which we have failed to reject, and the research hypothesis, which cannot be directly tested with statistics. Tests exist only for the null hypothesis. Thus, if we fail to reject the null hypothesis, we have an inconclusive result because there are two hypotheses that explain the difference.

Let us review the ***decision rules*** for the *z* test we have just considered:

    1. If the value of *z* that you computed is as extreme as:[3]

        1.96 or –1.96, declare the difference to be significant at the .05 level (i.e., reject the null hypothesis).

        2.58 or –2.58, declare the difference to be significant at the .01 level (i.e., reject the null hypothesis).

    2. If the value of *z* that you computed does not meet the first condition, do *not* declare the difference to be significant and do *not* reject the null hypothesis.

Let us apply the decision rules to several examples. In each example, a random sample has been drawn and tested. In each case, the population mean and standard deviation are known:

Example 1:
Researcher Smith obtained a value of *z* of 3.458 for the difference between two means. She chose the .05 level before starting her study.

---

[3]Select one of the two rules *before* examining the data.

1. Should she reject the null hypothesis?

   Yes, because 3.458 is more extreme than 1.96.

2. Should she declare the difference to be statistically significant?

   Yes.

Example 2:

Researcher Doe obtained a value of *z* of 1.786 for the difference between two means. He chose the .05 level before starting his study.

1. Should he reject the null hypothesis?

   No, because 1.786 is not as extreme as 1.96.

2. Should he declare the difference to be statistically significant?

   No.

Example 3:

Researcher Jones obtained a value of *z* of 2.966 for the difference between two means. She chose the .01 level before starting her study.

1. Should she reject the null hypothesis?

   Yes, because 2.966 is more extreme than 2.58.

2. Should she declare the difference to be statistically significant?

   Yes.

Example 4:

Researcher Daly obtained a value of *z* of −1.999 for the difference between two means. He chose the .01 level before starting his study.

1. Should he reject the null hypothesis?

   No, because −1.999 is not as extreme as −2.58.

2. Should he declare the difference to be statistically significant?

   No.

The *decision rules* that you have learned about in this section apply to what are known as *two-tailed tests*. These are usually appropriate and, thus, widely used. *Two-tailed tests* and *one-tailed tests* are defined and compared in the next section.

Terms to Review Before Attempting Worksheet 35

z test, decision rules

Worksheet 35: *z* Test for One Sample

> *Riddle*: According to Voltaire, what does the length of an argument tell us about who is right?

DIRECTIONS: To find the answer to the riddle, write the answer to each question in the space immediately below it. The word in parentheses in the solution section next to the answer to the first question is the first word in the answer to the riddle, the word beside the answer to the second question is the second word, and so on.

1. If the standard deviation of a population is 50.00 and a sample of 42 subjects is drawn at random, what is the value of the standard error of the mean?

2. In a study, the mean of a population equals 45.00, the mean of a random sample equals 48.00, and the standard error of the mean equals 2.540. What is the value of *z*?

3. In a study, the mean of a population equals 100.00, the mean of a random sample equals 90.00, and the standard error of the mean equals 4.110. What is the value of *z*?

4. What constants are used to determine if a *z* is an unlikely event at the .05 level?

5. What constants are used to determine if a result is significant at the .01 level?

Worksheet 35 (Continued)

6. For the type of study described in this section, should the null hypothesis be rejected at the .05 level if z equals 2.343?

7. For the type of study described in this section, should the null hypothesis be rejected at the .05 level if z equals 1.74?

8. "For the type of study described in this section, a value of z of 2.997 means that the null hypothesis should be rejected at the .01 level." Is this statement true or false?

9. "*Not rejecting the null hypothesis* is equivalent to *accepting the null hypothesis*." Is this statement true or false?

Solution section:

```
        true (are)   7.715 (a)   1.58 and −1.58 (fighting)   yes (both)

    1.181 (long)   6.480 (quarrel)   2.96 and −2.96 (nevertheless)

       1.59 (listening)   −2.433 (dispute)   2.58 and −2.58 (that)

  no (parties)   false (wrong)   1.96 and −1.96 (means)   0.243 (time)
```

Write the answer to the riddle here, putting one word on each line: _____ _____ _____ _____

_____ _____ _____ _____ _____

Section 36: One-Tailed Versus Two-Tailed Tests

In the previous section, you learned how to conduct a ***two-tailed*** z test. You probably recall that our decision rule at the .05 level was to reject the null hypothesis when the value of z is as extreme as 1.96 *or* –1.96. The following figure illustrates that the rule is based on the two tails of the distribution, each of which contain 2.5% of the area under the normal curve. (Together, they contain 5%.)

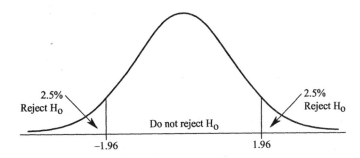

Figure 36.1. Two-tailed z test values for rejecting the null hypothesis at the .05 level.

In the example in the previous section, the population mean on the *SAT-Verbal* was compared with the mean of a random sample of students from the State of Disrepair (the 51st state). By using a two-tailed test, we were expressing our interest in one of *two* types of significant differences:

1. The sample mean is significantly lower than the population mean (as indicated by a z as extreme as –1.96.)

2. The sample mean is significantly higher than the population mean (as indicated by a z as extreme as 1.96.)

In other words, by using a two-tailed test, we were prepared to detect a difference in either direction even though the investigators' research hypothesis was only that the students in the State of Disrepair were significantly below the national average. But doesn't it make sense that if they turned out to be significantly *above* the national average, the investigators from the Department of Education of that state would want to be prepared to make this discovery? Had they selected a one-tailed test initially (and the rules of the game state that a two-tailed or one-tailed test must be selected initially before examining the data), they would have had to forego conducting a significance test of this interesting difference even if their students, on the average, had been vastly superior to the national average.

So if a ***two-tailed test*** provides more flexibility in examining the outcomes of a study, why would someone choose a ***one-tailed test***? Because a one-tailed test makes it easier to reject the null hypothesis—but in one and only one direction. The following figure illustrates that for the .05 level, we would use a critical value of −1.65 to conduct a one-tailed z test. The entire five percent of the area is in one tail. (Compare the two figures in this section.)[1]

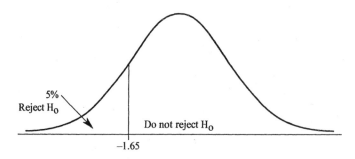

Figure 36.2. One-tailed z test value for rejecting the null hypothesis at the .05 level.

Notice that it is easier to reach −1.65 than −1.96 (the value for a two-tailed test).[2] In other words, −1.96 is a less likely event than −1.65 because it is

---

[1]The null hypothesis for the two-tailed test that we have been considering takes the form with which you are already familiar. It states that, in truth, the sample mean equals the population mean. The null hypothesis for a one-tailed test states that, in truth, the sample mean is equal to or greater than the population mean.

[2]For a one-tailed z test of whether the sample mean is significantly higher than the population mean, use a critical value of 1.65. For corresponding one-tailed tests at the .01 level, use the critical values of 2.33 and −2.33.

farther from the mean. We will be examining a number of other tests in this book. The emphasis will be on two-tailed tests for the following reasons:

1. Many consumers of research frown on a one-tailed test. They suspect that it may have been chosen only because it made it possible to report a significant difference and not because the underlying logic of the study justified a one-tailed test.

2. In most cases, it is difficult to justify a one-tailed test.[3] It can be justified only if you can convince your audience that there would be no interest in and no implications from a significant difference in a direction other than the one hypothesized in a directional research hypothesis. Usually, an astute consumer can imagine implications. Let us consider a couple of examples:

Example 1:

A researcher hypothesizes that subjects who take Vitamin E supplements will, after a period of years, have fewer wrinkles of the skin. This is a directional hypothesis. (See Section 33 to review directional and nondirectional hypotheses.) This researcher might argue that she is interested in significance in only one direction (fewer wrinkles in the experimental group than in the control group) and, therefore, plans to use a one-tailed test.

But suppose the unexpected happened and those who took the vitamin supplements had many more wrinkles than the control group. Would this be of interest? Would it have implications? Of course. People taking such supplements on their own might want to cut back their consumption of them as well as their consumption of foods rich in Vitamin E. Yet, by specifying a one-tailed test in advance, the researcher would be ethically bound *not* to switch and test for significance in the other direction. (Remember, the statistical rules of the game are that by opting for a lower standard for

---

[3]Some authors of statistics textbooks imply that a one-tailed test is justified whenever a researcher holds a directional research hypothesis (see Section 33). This is a much more liberal standard than suggested in this section.

rejecting the null hypothesis, one must agree in advance of seeing the data to test only in one direction.)

Example 2:

A researcher hypothesizes that a new computer-assisted program for teaching remedial math is superior to an existing method. This is a directional hypothesis and might be used to justify a one-tailed test. The reasoning might be that if the program is not superior to the existing method, the school district will not switch to the program; it will simply continue to use the existing method and ignore the program.

But suppose the unexpected happened and the students who used the computer program scored much lower than those who used the existing program. Wouldn't we want to know if this difference is statistically significant? I think so. The process of refining the educational process and keeping it up-to-date is an ongoing one. This process is accelerated if we not only know what procedures and programs are significantly better than traditional ones, but also which ones are significantly worse.

Should you ever use a one-tailed test? Yes, if you have a directional hypothesis *and* can convince yourself and your audience that a significant difference in the direction other than the one you hypothesized is of no interest.

Terms to Review Before Attempting Worksheet 36

one-tailed test, two-tailed test

Worksheet 36: One-Tailed Versus Two-Tailed Tests

Riddle: Early to bed and early to rise does what?

DIRECTIONS: To find the answer to the riddle, write the answer to each question in the space immediately below it. The word in parentheses in the solution section next to the answer to the first question is the first word in the answer to the riddle, the word beside the answer to the second question is the second word, and so on.

1. In a normal curve, what percentage of the area is above a z-score of 1.96?

2. If we are interested in whether a sample mean is either significantly higher or significantly lower than the population mean, should we use a "one-tailed" *or* "two-tailed test"?

3. If we were only interested in whether a sample mean is significantly higher than the population mean, should we use a "one-tailed" *or* "two-tailed test"?

4. In a normal curve, what percentage of the area is above a z-score of 1.65?

5. "If an investigator obtained a difference in the direction indicated by her research hypothesis, she would be more likely to be able to declare the difference to be significant if she used a one-tailed test than if she used a two-tailed test." Is this statement true or false?

Worksheet 36 (Continued)

6. "If a researcher declares a difference to be significant because the value of z is as extreme as 1.65, he is using a two-tailed test." Is this statement true or false?

7. If a researcher would be interested in a significant difference in either direction, should a one-tailed test be used?

Solution section:

yes (seeing)   1% (dawn)   no (dead)   false (socially)   true (but)
0.005% (leads)   5% (healthy)   2.5% (makes)   two-tailed (a)
one-tailed (person)   maybe (good)   0.5% (night)   reject it (are)

Write the answer to the riddle here, putting one word on each line: _____ _____ _____ _____
_____ _____ _____

Section 37: Introduction to the *t* Test

In this section, we will consider a frequently encountered problem: how to compare the means of two samples for statistical significance. Let us consider two examples:

Example 1:

An investigator wanted to determine whether there are differences between men and women voters in their attitudes toward welfare. Samples of men and women were drawn at random and administered an attitude scale to obtain a score for each subject. Means for the two samples were computed. Women had a mean of 40.00 (on a scale from 0 to 50, where 50 is the most favorable). Men had a mean of 35.00. The researcher wants to determine whether there is a significant difference between men and women. What accounts for the 5-point difference? One possible explanation is the *null hypothesis*, which states that there is no true difference between men and women—that the observed difference is due to sampling errors created by random sampling.

Example 1 illustrates that two means may be obtained from a ***survey***—a descriptive study in which a sample is assessed in order to draw inferences to its population.

Example 2:

A random sample of kittens is fed a vitamin supplement from birth to see if the supplement increases their visual acuity. Another random sample is fed a placebo that looks like the supplement but contains no vitamins. At the end of the study, both samples are tested for visual acuity and an average acuity score is calculated for each sample. The kittens that took the supplement scored 4 points higher on the average than the control group. What accounts for the

4-point difference? One possible explanation is the *null hypothesis*, which states that there is no true difference between the two samples of kittens—that the observed difference is due to sampling errors created by random sampling.

Example 2 illustrates that two means may be obtained from an ***experiment***—a study in which treatments are given in order to observe for their effects.

Surveys and experiments are very frequently conducted, and they often yield two means each, so you can see how important it is to be able to test the null hypothesis for the difference between two sample means.[1] We cannot use a *z* test, which was covered in Section 35, because we do not know the standard deviation of the population, which is required in conducting that test. Fortunately, about a hundred years ago, a statistician named William Gosset developed the ***t test*** for exactly the situations we are considering. As a test of the null hypothesis, it yields a probability that a given null hypothesis is correct. When the probability that it is correct is low—say .05 or 5% or less—we usually reject the null hypothesis.

The computational procedures for conducting *t* tests are covered in Sections 38 and 40. Before we get to that, let us consider what makes the *t* test work. In other words, what leads the *t* test to give us a low probability that the null hypothesis is correct? Here are the three basic factors:

1. The larger the samples, the less likely the difference between two means was created by sampling errors. You probably already intuitively knew that larger samples have less sampling error than smaller ones. Thus, when large samples are used, the *t* test is more likely to yield a probability low enough to allow us to reject the null hypothesis than when small samples are used.[2]

---

[1] Other types of studies also yield two sample means that are to be compared.
[2] You may recall that increasing sample size yields diminishing returns in terms of reducing errors. (See Section 27.)

2. The larger the difference between the two means, the less likely that the difference was created by sampling errors. Random sampling tends to create many small differences and few large ones. Thus, when large differences between means are obtained, the *t* test is more likely to yield a probability low enough to allow us to reject the null hypothesis than when small differences are obtained.

3. The smaller the variance among the subjects, the less likely that the difference between two means was created by sampling errors. To understand this, consider a population in which everyone is identical—they all look alike, think alike, and speak and act in unison. How many do you have to sample to get a good sample? Only one, because they are all the same. Thus, when there is no variation among subjects, it is not possible to have sampling errors. As the variation increases, sampling errors are more and more likely to occur.[3]

There are two types of *t* tests. One is for ***independent data*** (sometimes called *uncorrelated data*) and one is for ***dependent data*** (sometimes called *correlated data*). Examples 1 and 2 on pages 221 and 222 have independent data. Example 3 describes a study with dependent data.

Example 3:

In a study of visual acuity, same-sex siblings (two brothers or two sisters) were identified for a study. For each pair of siblings, a coin was tossed to determine which one received a vitamin supplement and which one received a placebo. Thus, in the control group, there is a subject who is a same-sex sibling of each subject in the experimental group.

---

[3]In the types of studies we are considering, we do not know the population standard deviation, which would indicate the amount of variation. The *t* test uses the standard deviations of the samples to estimate the variation of the population.

The means that results from the study in Example 3 are subject to less error than the means from Example 2. Remember that in Example 2, there was no matching or pairing of subjects before assignment to conditions. In Example 3, the matching of subjects assures us that the two groups are more similar than if just two independent samples were used. To the extent that genetics and gender are associated with visual acuity, the two groups in Example 3 will be more similar at the onset of the experiment than the two groups in Example 2.[4] The *t* test for dependent data takes this possible reduction of error into account. Thus, it is important to select the right *t* test.

Section 38 illustrates how to conduct a *t* test for independent data, Section 39 illustrates how to interpret the results of *t* tests in general, and Section 40 illustrates how to conduct it for dependent data. Those of you using computers to perform calculations may not need to master the calculations in Sections 38 and 40; however, you should examine them to learn more about how the *t* test works.

Terms to Review Before Attempting Worksheet 37

survey, experiment, *t* test, independent data, dependent data

---

[4]Ideally, we would like to conduct an experiment in which the two groups are initially *identical* in their visual acuity. This would make it more likely that any differences in acuity at the end of the experiment were due to the vitamin supplement and not to initial group differences.

Worksheet 37: Introduction to the *t* Test

> *Riddle*: According to Helen Rowland, why does a bachelor get tangled up with a lot of women?

DIRECTIONS: To find the answer to the riddle, write the answer to each question in the space immediately below it. The word in parentheses in the solution section next to the answer to the first question is the first word in the answer to the riddle, the word beside the answer to the second question is the second word, and so on.

1. Is a study in which treatments are given in order to observe for effects called a "survey" *or* an "experiment"?

2. "Two sample means can be obtained only from experiments." Is this statement true or false?

3. "The larger the difference between two means, the less likely that the difference was created by sampling errors." Is this statement true or false?

4. Other things being equal, will large samples or small samples be more likely to lead to rejection of the null hypothesis?

5. Other things being equal, will samples with little variance or samples with much variance be more likely to lead to rejection of the null hypothesis?

Worksheet 37 (Continued)

6. "A simple random sample of subjects was selected for the experimental group and another simple random sample of subjects was selected for the control group." Will this design result in "independent data" *or* "dependent data"?

7. "Subjects were paired according to their ability and then a coin was tossed for each pair to determine which subject was assigned to the experimental group; the remaining member of each pair was assigned to the control group." Will this design result in "independent data" *or* "dependent data"?

8. In general, is there usually more sampling error in independent data than in dependent data?

Solution section:

independent data (tied)   experiment (in)   survey (wedding)

false (order)   yes (one)   no (argument)   dependent data (to)

large samples (avoid)   small samples (can)   much variance (silly)

little variance (getting)   true (to)   *t* (wisely)   sampling error (at)

Write the answer to the riddle here, putting one word on each line: _____ _____ _____ _____
_____ _____ _____ _____

Section 38: Computation of *t* for Independent Data

As noted in the previous section, independent data are obtained when there is no matching or pairing of subjects across groups. In this section, we will first examine how to compute *t* for independent data and how to interpret it using the *t* table.

The formula for *t* is simple:

$$t = \frac{m_1 - m_2}{S_{Dm}}$$

Where:

m_1 is the mean of the group with the higher mean.
m_2 is the mean of the group with the lower mean.
S_{Dm} is the standard error of the difference between means.

The numerator of the formula is easy to understand. It is the difference between the two means. As you can see, the larger the difference, the larger the value of *t*.[1]

The denominator starts with the familiar symbol *S* (for standard deviation). The subscripts (*D* for difference and *m* for means) tell us that it is the standard deviation of the difference between means; this standard deviation is called the *standard error of the difference between means*. You probably recall that in Section 31 we calculated the standard error of a single mean in order to interpret it in light of sampling errors. What we are interpreting in this section is the *difference between two means*; thus, we need the standard deviation of this difference. Once we have it, we can use the technique that should be familiar to you by now—determining whether an event is unlikely to occur by chance in light of the number of standard deviations some statistic is from the mean of the distribution. In this case, we want to know whether the difference between two means is an unlikely event. If it is unlikely (for example, likely to occur less than 5 times in 100 due to chance alone), then we will declare the

---

[1]As you will see later in this section, the larger the value of *t*, the more likely it is that the null hypothesis can be rejected.

difference to be statistically significant; that is, unlikely to be the result of random errors.

It is impractical to directly obtain the S_{Dm} for a given *t* test.[2] So what we do is estimate it given what we know about the sample size and the variance of the samples (remember that the *variance* is simply the square of the standard deviation, whose symbol is s^2), using this formula:

$$S_{Dm} = \sqrt{\left[\frac{(n_1 - 1)(s_1^2) + (n_2 - 1)(s_2^2)}{n_1 + n_2 - 2}\right]\left[\frac{1}{n_1} + \frac{1}{n_2}\right]}$$

Where:

n_1 is the number of cases in Group 1.
n_2 is the number of cases in Group 2.
s_1 is the standard deviation of Group 1 (which will be squared).
s_2 is the standard deviation of Group 2 (which will be squared).
NOTE: When obtaining the standard deviations, you should use the second formula for the standard deviation given in Appendix A.

It is really quite easy to use. Here is an example:

Example 1:

For Group 1, $m_1 = 24.000$, $s_1 = 1.500$, and $n_1 = 12$
For Group 2, $m_2 = 22.000$, $s_2 = 1.400$, and $n_2 = 11$

$$S_{Dm} \sqrt{\left[\frac{(12 - 1)(1.500^2) + (11 - 1)(1.400^2)}{12 + 11 - 2}\right]\left[\frac{1}{12} + \frac{1}{11}\right]}$$

$$\sqrt{\left[\frac{(11)(2.250) + (10)(1.960)}{21}\right][.083 + .091]}$$

---

[2]If we were to draw an infinitely large number of two samples at random and each time compute the two means of the samples, then compute all the differences between the means, and finally compute the standard deviation of all the differences, we would have the standard error of the difference between means (S_{Dm}) for a given study. We could then compare the difference between means for a given study with this distribution of many differences between means in order to determine whether the difference in the study is unlikely to occur by chance. This is, of course, impractical.

$$= \sqrt{\left[\frac{24.750 + 19.600}{21}\right][.174]}$$

$$= \sqrt{\left[\frac{44.350}{21}\right][.174]} = \sqrt{[2.112][.174]} = \sqrt{.367} = .606$$

Thus, for our example, the value of S_{DM} equals .606. We substitute it into the formula for *t* as follows:

$$t = \frac{m_1 - m_2}{S_{Dm}} = \frac{24.000 - 22.000}{.606} = \frac{2.000}{.606} = 3.300$$

We call the result the ***observed value of t***. Thus, in this example, we observed a value of 3.300. To evaluate its meaning, we need to take account of the number of cases that underlie it, using this formula for the ***degrees of freedom (df)***:[3]

$$df = n_1 + n_2 - 2$$

Where:

n_1 is the number of cases in Group 1.

n_2 is the number of cases in Group 2.

2 is a constant for this type of problem. (Use it whenever you are conducting a *t* test on two means for independent data.)

For our example:

$$df = 12 + 11 - 2 = 23 - 2 = 21$$

---

[3]As you can see by studying its formula, degrees of freedom are directly related to the total number of cases. The name intrigues many students, although its mathematical derivation is not of use to students of applied statistics. For those who are interested, it comes from the fact that all but two of the cases can vary (take on any value) and the same means can be obtained if the remaining two are adjusted to appropriate values. Thus, all but two cases are "free to vary."

You know from previous sections that if we had obtained a value of *z* as extreme as 1.96, the result would be declared to be an unlikely event. That is, it would be declared to be unlikely to occur by chance because the odds are less than .05 that this is a chance deviation in a normal distribution. However, this is a *t* test, which is based on the fact that the underlying distributions are not normal in shape when the sample size is small. Thus, instead of using constants such as 1.96 to evaluate the value of *t*, we use the appropriate **critical value of t** found in the *t* table in Table 4 near the end of this book.[4]

Examining Table 4, we find that for an infinite number of degrees of freedom (at the bottom of the table), the familiar 1.96 is the critical value. However, in our example, the degrees of freedom equal 21. Look up 21 in the first column, then look to the right to the .05 column. There you find a *critical value* of 2.080. We have found that for 21 degrees of freedom, only values as extreme as 2.080 are unlikely events at the .05 level. Our *observed value* is 3.300. Is this an unlikely event? Yes. Thus, we can reject the null hypothesis and declare the result to be statistically significant at the .05 level.[5]

Is our *observed value* of *t* of 3.300 an unlikely event at the .01 level? Yes, because the *observed value* of 3.300 exceeds the *critical value* for the .01 level for 21 degrees of freedom, which is 2.831. Thus, we can reject the null hypothesis and declare the result to be statistically significant at the .01 level.

Is our *observed value* of *t* of 3.300 an unlikely event at the .001 level? No, because the *observed value* of 3.300 does *not* exceed the *critical value* for the .001 level for 21 degrees of freedom, which is 3.819. Thus, we *cannot* reject the null hypothesis at the .001 level and cannot declare the result to be statistically significant at this level.

In review, here is how to use the *t* table (Table 4):

---

[4]Table 4 near the end of this book is for a two-tailed *t* test. For a one-tailed test, use Table 5 near the end of this book. See Section 36 to review the differences between one-tailed and two-tailed tests.

[5]This probability is correct if the assumptions underlying the *t* test are met: (1) random samples from populations with normal distributions are drawn, and (2) the variances of the two populations are similar. Mild violations of these assumptions have little effect on the probabilities. When there is serious doubt as to the tenability of the assumptions, use larger samples (say, 20 or more) to obtain relatively accurate probabilities in the face of violations of the assumptions.

1. Look up the degrees of freedom for your problem in the first column.

2. Move to the right to the significance level you are interested in (.05, .01, or .001) to determine the *critical value*.[6]

3. If the *observed value is greater than the critical value*,[7] reject the null hypothesis; otherwise, do not reject it.

You may have noticed that not all possible values of degrees of freedom (*df*) are shown in Table 4. For example, suppose you conducted a study in which there were 32 degrees of freedom. Table 4 shows the critical values of *t* for 30 *df* (2.042 for the .05 level) and 40 *df* (2.021 for the .05 level) but not for exactly 32 degrees of freedom. For this type of situation, use the *lower value of df*. In the case under consideration, for a *df* of 32, use the critical value associated with 30 degrees of freedom.

In the next section, we will examine how various authors report values of *t* and how to interpret their reports.

Terms to Review Before Attempting Worksheet 38

observed value of *t*, critical value of *t*, degrees of freedom, *df*

---

[6]You should select the level you are interested in before examining any of the data.

[7]In the unlikely event that the observed value is exactly equal to the critical value, reject the null hypothesis.

Worksheet 38: Computation of *t* for Independent Data

> *Riddle*: According to Henry Wadsworth Longfellow, it takes less time to do a thing right than...

DIRECTIONS: To find the answer to the riddle, write the answer to each question in the space immediately below it. The word in parentheses in the solution section next to the answer to the first question is the first word in the answer to the riddle, the word beside the answer to the second question is the second word, and so on.

Questions 1 through 3 refer to the information in this box.

> For Group 1, $m = 30.000$, $s = 2.100$, $n = 14$
> For Group 2, $m = 25.000$, $s = 1.900$, $n = 15$

1. What is the value of the standard error of the difference between means? (The answer in the solution section was obtained by rounding at each step to three decimal places. Allow for minor differences if you use a different procedure.)

2. What is the observed value of *t*? (Use the value of S_{Dm} in the solution section for Question 1.)

3. Based on a two-tailed test, is the difference statistically significant at the .05 level?

4. What is the critical value of *t* for a two-tailed test at the .05 level for a problem in which there are 6 degrees of freedom?

Worksheet 38 (Continued)

5. For a problem in which the observed value of $t = 1.999$ with 21 degrees of freedom (two-tailed), may the null hypothesis be rejected at the .05 level?

6. "Based on the information in Question 5, the difference between the means is statistically significant at the .05 level." Is this statement true or false?

7. "For a problem in which the observed value of $t = 2.500$ with 26 degrees of freedom (two-tailed), the difference is significant at the .05 level but not at the .01 level." Is this statement true or false?

8. Using the information in Question 7, what decision should be made about the null hypothesis at the .05 level?

9. If $n = 30$ for one group and $n = 32$ for another group, what is the value of df for a t test on independent data?

Solution section:

false (you)	2.348 (mistake)	.872 (lengthy)	2.447 (explain)	true (did)
.742 (it)	6.739 (does)	no (why)	3.707 (always)	reject it (it) yes (to)
do not reject it (helpless)	62 (being)	60 (wrong)	61 (difficult)	

Worksheet 38 (Continued)

Write the answer to the riddle here, putting one word on each line: _____ _____ _____ _____
_____ _____ _____ _____ _____

"Now I have zero degrees of freedom."

Section 39: Reporting the Results of *t* Tests

We are considering the use of the *t* test to test the difference between two sample means for significance. Obviously, you should report the values of the means before reporting the results of the test on them. In addition, you should report the values of the standard deviations and the number of cases in each group. This may be done within the context of a sentence or in a table. Table 39.1 shows a typical table.

Table 39.1
Means and Standard Deviations

	m	*s*	*n*
Group A	2.50	1.87	6
Group B	6.00	1.89	6

The samples that formed Groups A and B were drawn at random. The null hypothesis states that the 3.50-point difference ($6.00 - 2.50 = 3.50$) between the means of 2.50 and 6.00 are the result of sampling errors (i.e., errors resulting from random sampling) and that the true difference in the population is zero.[1]

The results of the *t* test may be described in several ways. Here are some examples for the results in Table 39.1:

Example 1:
The difference between the means is statistically significant
($t = 3.22$, *df* = 10, $p < .01$, two-tailed test).

To the sophisticated reader, the results in Example 1 indicate that the null hypothesis has been rejected because "statistically significant" is synonymous with "rejecting the null hypothesis."

---

[1]This statement of the null hypothesis is for a two-tailed test. See Section 36 to review the difference between one-tailed and two-tailed tests.

Example 2:

The difference between the means is significant at the .01 level
($t = 3.22$, $df = 10$, two-tailed test).

In Example 2, the author has indicated that significance was obtained at the .01 level. This tells us that p was equal to or less than .01. Thus, the null hypothesis was rejected.

Example 3:

The null hypothesis was rejected at the .01 level ($t = 3.22$, $df =$ 10, two-tailed test).

From Example 3, we know that the difference is statistically significant since rejecting the null hypothesis is the same as declaring statistical significance.

Any of the three forms shown above is acceptable. Authors of journal articles tend to talk about differences as being either "statistically significant" or "statistically insignificant" and seldom mention the null hypothesis. In theses and dissertations, explicit references to the null hypothesis are more common.

When you use the word "significant" when reporting the results of significance tests, you should always modify it with the adjective "statistically." This is because a result may be **statistically significant** but not of any **practical significance**. For example, suppose you found a statistically significant difference of 2 points in favor of a computer-assisted approach over a traditional lecture/textbook approach. While it is statistically significant, it may not be of practical significance if the school district has to invest sizable amounts of money to buy new hardware and software; in other words, the cost of the difference may be too great in light of the absolute size of the benefit.

Let us consider how to report a difference that is not significant. Table 39.2 presents some results. Examples 4 through 6 show some ways to express the results of the *t* test.

Table 39.2
Means and Standard Deviations

	m	*s*	*n*
Group X	8.14	2.19	7
Group Y	5.71	2.81	7

Example 4:

The difference between the means is not statistically significant ($t = 1.80$, $df = 12$, $p > .05$, two-tailed test).

The fact that *p* is *greater than* (>) .05 suggests that we should not reject the null hypothesis and not declare statistical significance.

Example 5:

For the difference between the means, $t = 1.80$ ($df = 12$, *n.s.*, two-tailed test).

The author of Example 5 has used the abbreviation *n.s.* to tell us that she has declared the difference to be not significant. Because we are not given a specific probability level, most readers will assume that it was not significant at the .05 level—the most liberal of the widely used levels. Example 4 is preferable to Example 5 because Example 4 indicates the specific probability level in question.

Example 6:

The null hypothesis for the difference between the means was not rejected at the .05 level ($t = 1.80$, $df = 12$, two-tailed test).

While reading journal articles, theses, and dissertations, you will find variations in the exact words used to describe the results of *t* tests. The examples in this section show you models that you might use in your own writing.

Terms to Review Before Attempting Worksheet 39

statistically significant, practical significance

**Many students actually look forward
to Mr. Atwadder's statistics tests.**

Worksheet 39: Reporting the Results of *t* Tests

Riddle: According to Ogden Nash, for a successful marriage, whenever you are wrong, admit it. What else should you do?

DIRECTIONS: To find the answer to the riddle, write the answer to each question in the space immediately below it. The word in parentheses in the solution section next to the answer to the first question is the first word in the answer to the riddle, the word beside the answer to the second question is the second word, and so on.

Box A

> The difference between the means is statistically significant ($t = 2.12$, $df = 26$, $p < .05$, two-tailed test).

1. For Box A, was the null hypothesis rejected at the .05 level?

Box B

> The null hypothesis was rejected at the .05 level ($t = 3.145$, $df = 6$, two-tailed test).

2. "For Box B, the difference is statistically significant at the .05 level." Is this statement true or false?

Worksheet 39 (Continued)

3. Is the statement in "Box A" *or* the statement in "Box B" in the form that is more often used in journals?

Box C

> The difference between the means is not statistically significant ($t = 1.72$, $df = 25$, two-tailed test).

4. What important type of information is missing in Box C?

5. For Box C, has the author rejected the null hypothesis?

6. The abbreviation *n.s.* stands for what two words?

Solution section:

true (you) probability level (right) Box B (cowardice) no (shut)		
null significance (last) not significant (up) Box A (are) yes (whenever)		
Box C (running) false (bloody) standard deviation (cry)		

Write the answer to the riddle here, putting one word on each line: _____ _____ _____ _____,
_____ _____

Section 40: Computation of *t* for Dependent Data

Our goal is to test the null hypothesis for the difference between two means when dependent data are being analyzed. Dependent data are obtained when each score in one set of scores is paired with a score in another set. See Section 37 to review the difference between dependent and independent data.

The formula for *t* looks almost the same as the formula for a *t* test for independent data presented in Section 38. It is:

$$t = \frac{m_1 - m_2}{S_{mD}}$$

Where:

        m_1 is the mean of the group with the higher mean.
        m_2 is the mean of the group with the lower mean.
        S_{mD} is the standard error of the mean difference.

The numerator is the difference between the two means. As you can see, the larger the difference, the larger the value of *t*.

The denominator starts with the familiar symbol *S* (for standard deviation). The subscripts (*m* and *D*) tell us that it is the standard deviation of the mean difference. This is the formula for S_{mD}:

$$S_{mD} = \sqrt{\frac{\Sigma D^2 - (\Sigma D)^2 \div n}{n(n-1)}}$$

Where:

        D is the difference between a pair of means.
        n is the number of *pairs* of cases.

To use this formula, first list the two sets of scores, making sure that two paired scores are on each line. The following table illustrates this. In Column 1 are the names of the pairs. For example, each pair could be two identical twins, one of whom was randomly assigned to the experimental group and the

Table 40.1
Scores for Pairs of Subjects

	Experimental (X_1)	Control (X_2)	D	D^2
Pair A	8	5	3	9
Pair B	12	10	2	4
Pair C	10	11	−1	1
Pair D	9	6	3	9
Pair E	18	15	3	9
Pair F	11	7	4	16
Pair G	8	2	6	36
SUMS	$\Sigma X_1 = 76$	$\Sigma X_2 = 56$	$\Sigma D = 20$	$\Sigma D^2 = 84$

other assigned to the control group.[1] The difference between each pair of scores was computed and entered in the column under D. Then the differences were squared and entered under the column for D^2. Then, substituting into the formula for the standard error of the mean difference, we obtain:

$$S_{mD} = \sqrt{\frac{\Sigma D^2 - (\Sigma D)^2 \div n}{n(n-1)}}$$

(Notice that in the numerator, we divide before subtracting in the subsequent steps.)

$$\sqrt{\frac{84 - (20)^2 \div 7}{7(7-1)}}$$

$$\sqrt{\frac{84 - 400 \div 7}{7(6)}}$$

$$\sqrt{\frac{84 - 57.143}{42}} = \sqrt{\frac{26.857}{42}} = \sqrt{.639} = .799$$

Before we can solve for t, we must first compute the two means: $m_1 = 76/7 = 10.857$ and $m_2 = 56/7 = 8.000$. We now have the three values required by the formula for t. These have been substituted into the following formula:

---

[1] The members of the pairs do not have to be related. For example, pairs could have been formed by matching the subjects on the basis of their scores on an achievement test.

$$t = \frac{m_1 - m_2}{S_{mD}} = \frac{10.857 - 8.000}{.799} = \frac{2.857}{.799} = 3.576 = 3.58$$

In order to evaluate our observed value of *t* using the critical values of *t* in Table 4, we must first compute the degrees of freedom (*df*).[2] For dependent data, the formula is:

$$df = n - 1$$

Where:

n is the number of *pairs* of scores.

In this example, there are 14 scores but only 7 pairs. Therefore:

$$df = 7 - 1 = 6$$

Examination of Table 4 for 6 degrees of freedom reveals that the **critical value** for the .05 level is 2.447. Because our **observed value** of *t* (3.58) exceeds the critical value, the difference between the means is significant at the .05 level and the null hypothesis may be rejected at this level.

Table 4 also reveals that the critical value for the .01 level is 3.707. Because our observed value of *t* (3.58) does *not* exceed the critical value, the difference between the means is *not* significant at the .01 level and the null hypothesis may *not* be rejected at this level.[3]

Keep in mind that you should select the level at which you wish to test before examining any of the data. In this example, if you had selected the .05 level, you would have been allowed to declare the result to be significant; if you had selected the .01 level, you would not have been allowed to do so.

---

[2]For a one-tailed test, use the critical values in Table 5. See Section 36 to review the difference between a one-tailed and a two-tailed test.
[3]In the unlikely event that the observed value is exactly equal to the critical value, reject the null hypothesis.

In the next section, we will examine how to analyze the differences among two *or more* means. As you can see, the formula for *t* only allows you to enter the values for two means—limiting the usefulness of the *t* test.

Terms to Review Before Attempting Worksheet 40

critical value, observed value

"The committee met to approve your proposed data analysis for your thesis. But first we had to approve the approval, providing everyone agreed to disagree to approve the agreement, which approved the approval agreement. After that, things got complicated."

Worksheet 40: Computation of *t* for Dependent Data

> *Riddle*: According to Mark Twain, how do we know
> that most writers regard truth as valuable?

DIRECTIONS: To find the answer to the riddle, write the answer to each question in the space immediately below it. The word in parentheses in the solution section next to the answer to the first question is the first word in the answer to the riddle, the word beside the answer to the second question is the second word, and so on.

The questions refer to the information in this box.

	Experimental	Control
Pair A	8	5
Pair B	4	1
Pair C	3	4
Pair D	8	5
Pair E	9	5
Pair F	9	6

1. To three decimal places, what is the value of the difference between the two means?

2. To three decimal places, what is the value of the standard error of the mean difference? (The answer in the solution section was obtained by rounding at each step to three decimal places. Allow for minor differences if you use a different procedure.)

Worksheet 40 (Continued)

3. To two decimal places, what is the observed value of t?

4. What is the critical value of t at the .05 level for this problem? (Use Table 4.)

5. Is the difference between the means significant at the .05 level?

6. "The null hypothesis may be rejected at the .05 level." Is this statement true or false?

7. Is the difference between the means significant at the .01 level?

8. "The null hypothesis may be rejected at the .01 level." Is this statement true or false?

Solution section:

6.833 (as)   15.000 (lie)   false (use)   2.500 (because)   3.48 (are)   true (in)

53.00 (never)   2.571 (most)   yes (economical)   .719 (they)

37.500 (waver)   no (its)   .05 (argument)   .01 (discourse)   .001 (been)

Worksheet 40 (Continued)

Write the answer to the riddle here, putting one word on each line: _____ _____ _____ _____ _____ _____ _____ _____

"For my statistical experiment, I'm going to drive downtown, get a great parking spot, and then I'm going to count how many people ask me if I'm leaving."

Notes:

Section 41: Introduction to Analysis of Variance

In Sections 37 through 40, you learned about the *t* test, which tests the null hypothesis regarding the difference between *two* means. A closely related test is analysis of variance (***ANOVA***), which is sometimes informally called the *F* test. ANOVA is used to test the difference(s) among *two or more means*.

First, ANOVA can be used to test the difference between two means. To do so, we calculate a statistic called *F*, calculate the degrees of freedom for *F*, and evaluate *F* using a table of critical values of *F*. (The computational procedures are described in the next section.) After we have done this, the resulting *probability* will be the same as the probability we would have obtained using a *t* test. However, the value of *F* will not be the same as the value of *t*. In fact, for a given set of data yielding two means to be compared, $F = t^2$. Since the end result of primary interest in significance testing is the probability, if you have already mastered the *t* test, you do not need to learn how to use ANOVA to test the difference between two means. However, ANOVA can also be used to test the differences among more than two means in a single test, which cannot be done with a *t* test.[1]

Consider Example 1:

Example 1:
A new drug for treating migraine headaches was tested on three groups selected at random. The first group received 250 milligrams, the second received 100 milligrams, and the third received a placebo (an inert substance). The average reported pain for the three groups (on a scale from 0 to 20, with 20 representing the most pain) was determined by calculating means. The means for the groups were:[2]

---

[1]The assumptions underlying ANOVA are that the variances of the groups are similar (i.e., homogeneous), that the groups are independent, and that the subjects were selected at random.
[2]When reporting means, you should also report the associated values of the standard deviations.

Group 1: $m = 1.78$

Group 2: $m = 3.98$

Group 3: $m = 12.88$

As you can see in Example 1, there are three differences among means: (1) the difference between Groups 1 and 2, (2) the difference between Groups 1 and 3, and (3) the difference between Groups 2 and 3. Instead of running three separate t tests,[3] we can run a single ANOVA to test the significance of this *set of differences*. There are two ways the results of Example 1 can be reported. Example 2 shows one of these:

Example 2:
The difference between the means was statistically significant at the .01 level ($F = 58.769$, $df = 2, 36$).

Note that the method of reporting in Example 2 is similar to that for reporting the results of a t test.[4] This result tells us that there is a significant difference with $p < .01$. Thus, the null hypothesis may be rejected at the .01 level. The null hypothesis for this test says that the *set of three differences* was created at random. By rejecting the null hypothesis, we are rejecting the notion that *one or more* of the differences were created at random. Notice that the test does not tell us which of the three differences is responsible for the rejection of the null hypothesis. It could be that only one or two of the three differences were responsible for the significance. Procedures for determining which individual differences are significant are described in Sections 43 and 44.

Example 3 shows another way that the results of an ANOVA are commonly reported in journals. It is called an *ANOVA table*. In the table, you see the values of F, df, and p that were reported in Example 2. You also see the values of the *sum of squares* and the *mean square*—these are intermediate

---

[3]It would be inappropriate to run three separate t tests without an adjustment in the standard probabilities for t. Alternatives to doing this are presented in Sections 43 and 44.

[4]You probably noticed that there are two values reported as degrees of freedom for an ANOVA. The next section indicates how to obtain them and use them in evaluating F.

values obtained in the calculation of F. (For example, if you divide the mean square of 315.592 by the mean square of 5.370, you will obtain F.) Procedures for calculating these intermediate values are described in Section 42. For the typical consumer of research, however, these values are of little interest.[5]

Example 3:

Table 41.1
Analysis of Variance Table for the Data in Example 1

Source of Variation	df	Sum of Squares	Mean Square	F
Between Groups	2	631.185	315.592	58.769*
Within Groups	36	193.320	5.370	
Total	38	824.505		

*$p < .01$

Notice that the probability in Example 3 is given in a footnote, which is common. However, sometimes it will be given in the table and sometimes it will be given in the text that describes the table.

The technique we are considering can be generalized to a larger number of means. Consider Example 4:

Example 4:

Four methods of teaching computer literacy were used in an experiment, which resulted in four means. This produced these six differences:

1. The difference between Methods 1 and 2.

2. The difference between Methods 1 and 3.

3. The difference between Methods 1 and 4.

4. The difference between Methods 2 and 3.

---

[5]Those with advanced training in statistics can use these intermediate values to enhance their interpretation of the data. The typical reader is interested in whether or not the null hypothesis has been rejected, which is indicated by the value of p.

5. The difference between Methods 2 and 4.

6. The difference between Methods 3 and 4.

A single ANOVA can determine whether the null hypothesis for this entire set of six differences should be rejected. If the result is not significant, the researcher is done. If the result is significant, he or she may test to see which of the six differences are significant by using the techniques described in Sections 43 and 44.

The examples we have been considering are examples of what is known as a *one-way ANOVA* (also known as a *single-factor ANOVA*). This term is derived from the fact that subjects were classified *one* way. In Example 1, they were classified only according to the drug group to which they were assigned. In Example 4, they were classified only according to the method of instruction to which they were exposed. In Section 45, you will be introduced to *two-way ANOVA* (also known as a *two-factor ANOVA*) in which each subject is classified in two ways such as (1) which drug group they were assigned to and (2) whether he/she is male or female. A two-way ANOVA permits us to answer questions that are potentially more interesting.

In the next section, you will learn the computational techniques for a single-factor ANOVA. If you are using a computer to perform calculations, you may not need to master the section, but you should examine it carefully to obtain a better understanding of ANOVA.

Terms to Review Before Attempting Worksheet 41

ANOVA, one-way ANOVA, two-way ANOVA

Worksheet 41: Introduction to Analysis of Variance

> *Riddle*: According to Ingrid Bergman, a kiss is a lovely trick designed by nature to do what?

DIRECTIONS: To find the answer to the riddle, write the answer to each question in the space immediately below it. The word in parentheses in the solution section next to the answer to the first question is the first word in the answer to the riddle, the word beside the answer to the second question is the second word, and so on.

1. ANOVA can be used to test the differences among how many means?

2. For two means, will ANOVA and the t test yield the same probability?

3. In a comparison of three means, there are three differences. If F is significant, does this mean that all three differences are significant?

4. As a result of an ANOVA, this conclusion was drawn: The difference between the means was significant at the .05 level. What decision has been made about the null hypothesis?

5. For the typical consumer of research, what statistic in an ANOVA table is of greatest interest?

Worksheet 41 (Continued)

6. For four means in a one-way analysis of variance, how many differences among means result?

7. Does Example 1 in this section illustrate a "one-way ANOVA" *or* a "two-way ANOVA"?

Solution section:

> one-way (superfluous)   two-factor (love) F (affection) df (never)
>
> reject it (when)   do not reject it (only)   two or more (to)   yes (stop)
>
> no (speech)   4 (always)   6 (become) p (words)   accept it (vision)

Write the answer to the riddle here, putting one word on each line: _____ _____ _____ _____ _____ _____ _____

Section 42: Computations for a One-Way ANOVA

In the example that follows, subjects were classified according to which of three groups they were assigned to in an experiment: Group A received massive amounts of praise, Group B received moderate amounts of praise, and Group C received no praise for correct answers to math problems. Their scores on a posttest are shown in the first three columns of Table 42.1. At the bottom

Table 42.1
Scores for Three Groups on a Posttest and Related Statistics

Col. 1	Col. 2	Col. 3	Col. 4	Col. 5	Col. 6
Group A	Group B	Group C	A^2	B^2	C^2
7	4	3	49	16	9
6	6	2	36	36	4
5	4	1	25	16	1
8	7	3	64	49	9
3	5	4	9	25	16
7	7	1	49	49	1
$\sum X_A = 36$	$\sum X_B = 33$	$\sum X_C = 14$	$\sum X_A^2 = 232$	$\sum X_B^2 = 191$	$\sum X_C^2 = 40$
$m = 6.000$	$m = 5.500$	$m = 2.333$			

of these columns are the sums of the scores and the associated means (e.g., for Column 1, 36/6 = 6.000). In Column 4 are the squared scores for Group A, in Column 5 are the squared scores for Group B, and in Column 6 are the squares of the scores for Group C. The sums of the squared scores are shown at the bottom of Columns 4 through 6.

In order to test the significance of the differences among the three means, we need to compute the value of F and, using the associated values of the degrees of freedom, compare our computed value of F with the critical values of F in Table 6 (for the .05 level) or Table 7 (for the .01 level) in order to determine significance.[1] To get the value of F, follow these steps:

---

[1]This procedure is similar to the one used with the t test. Remember that we computed the value of t and the associated degrees of freedom and then compared the observed value of t with the critical values in Table 4.

Step 1: Set up a table such as Table 42.1; in this table there are three sets of scores, resulting in six columns. The number of columns will depend on how many groups you have. If you have only two sets of scores, there will be four columns; if you have four sets of scores, there will be eight columns, etc.

Step 2: Compute the ***total sum of squares*** (SS_T). This is the formula and its solution for the example under consideration:

$$SS_T = \Sigma X^2 - \frac{(\Sigma X_T)^2}{n}$$

Where:

ΣX^2 is the sum of the squared scores for each group (in this case, 232, 191, and 40).

ΣX_T is the sum of *all* the scores (in this case, the sum of all 18 scores, which can be obtained by adding the sums of the three sets of scores: $36 + 33 + 14 = 83$).

n is the total number of subjects (in this case, 18).

$$SS_T = 232 + 191 + 40 - \frac{83^2}{18} = \mathbf{80.278}$$

Step 3: Compute the ***between groups sum of squares*** (SS_b). This is a measure of the variation between the groups; it is needed in order to compute the value of F. The formula and its solution for the problem under consideration are shown here:

$$SS_b = \Sigma \frac{(\Sigma X)^2}{n} - \frac{(\Sigma X_T)^2}{n}$$

Where:

ΣX is the sum of the scores of each group (in this case, 36, 33, and 14, which will be used separately).

ΣX_T is the sum of *all* of the scores (in this case, the sum of all 18 scores, which is $36 + 33 + 14 = 83$).

n is the number of subjects in each group (in this case, 6).

N is the total number of subjects (in this case, 18).

$$SS_b = \frac{36^2}{6} + \frac{33^2}{6} + \frac{14^2}{6} - \frac{83^2}{18}$$

$$= 216.000 + 181.500 + 32.667 - 382.722$$

$$= 430.167 - 382.722 = \mathbf{47.445}$$

Step 4: Compute the ***within groups sum of squares*** (SS_w). This is an estimate of the amount of variation within groups.[2] We can obtain it by subtracting the SS_b from the SS_T, as shown here:

$$SS_w = SS_T - SS_b = 80.278 - 47.445 = \mathbf{32.833}$$

Let us pause and enter the values that we have calculated up to this point into an ANOVA table, which helps us organize and keep track of our statistics. Here is the table with the values computed up to this point.[3]

Table 42.2
ANOVA Table with Sum of Squares

Source of Variation	df	Sum of Squares	Mean Square	F
Between Groups	?	47.445	?	?*
Within Groups	?	32.833	?	
Total	?	80.278		

*p = ?

Step 5: Compute the *df* for between groups (df_b), which is the number of groups minus one:

$$df_b = n - 1 = 3 - 1 = \mathbf{2}$$

---

[2]The statistic F is a comparison of the variation within groups (which is the natural amount of variation among subjects) with the variation between groups (which in this case may have been caused by the experimental treatments). If the variation between is sufficiently greater than the variation within, the difference will be declared to be statistically significant.
[3]ANOVA tables were introduced in the previous section.

Step 6: Compute the df for the total (df_T), which is the total number of all subjects minus one:

$$df_T = N - 1 = 18 - 1 = \mathbf{17}$$

Step 7: Compute the df for within groups (df_w), which can be obtained by subtracting the df_b (for between) from the df_T for the total:

$$df_w = df_T - df_b = 17 - 2 = \mathbf{15}$$

Before proceeding, let us enter the values of the degrees of freedom into our ANOVA table:

Table 42.3
ANOVA Table with Sum of Squares and df

Source of Variation	df	Sum of Squares	Mean Square	F
Between Groups	2	47.445	?	?*
Within Groups	15	32.833	?	
Total	17	80.278		

*p = ?

Step 8: Compute the mean squares (*MS*). This is accomplished by dividing each sum of squares by its associated *df*. Thus, the **between groups mean square** (MS_b) is 47.445/2 = **23.723**; the **within groups mean square** (MS_w) is 32.833/15 = **2.189**.

Step 9: Compute the value of F by dividing the mean square for between groups by the mean square for within groups, as shown here:

$$F = \frac{MS_b}{MS_w} = \frac{23.723}{2.189} = 10.837$$

Let us enter the results of Steps 8 and 9 into our ANOVA table:

Table 42.4
ANOVA Table with Sum of Squares, df, Mean Squares, and F

Source of Variation	df	Sum of Squares	Mean Square	F
Between Groups	2	47.445	23.723	10.837*
Within Groups	15	32.833	2.189	
Total	17	80.278		

*$p = ?$

At this point, all we are missing is the value of p. To obtain this, we must compare our observed value of F (10.837) with the appropriate critical value in Table 6 (for the .05 level) or Table 7 (for the .01 level). Let us assume that we decided before examining the data to use the .05 level; thus, we should examine Table 6.[4]

The columns in Table 6 are labeled with degrees of freedom associated with the "Between Groups Degrees of Freedom" (df_b); as you can see in Table 42.4, this is 2. Thus, we should look at the column labeled "2." The rows are labeled with "Within Groups Degrees of Freedom"—in this case, it is 15. Thus, we go to where the column labeled "2" meets the row labeled "15." There we find a ***critical value of F*** of 3.68.

Here is the decision rule:

If the observed value of F (in this case, 10.837) is greater than the critical value (in this case, 3.68),[5] reject the null hypothesis; otherwise, do not reject it.

---

[4]Table 7 is read in the same way as Table 6.
[5]In the unlikely event that the observed value is exactly equal to the critical value, reject the null hypothesis.

Using the decision rule, we reject the null hypothesis at the .05 level and report that $p < .05$. Thus, the set of differences among the three means is statistically significant at the .05 level.

Table 42.5 is an ANOVA table with all the values:

Table 42.5
Completed ANOVA Table

Source of Variation	df	Sum of Squares	Mean Square	F
Between Groups	2	47.445	23.723	10.837*
Within Groups	15	32.833	2.189	
Total	17	80.278		

*$p < .05$

Although we now know that the set of differences among the three means (6.000, 5.500, and 2.333) is statistically significant at the .05 level, we do not know which individual pairs of differences (Group A vs. B, A vs. C, or B vs. C) are significant. A procedure for determining this is described in the next section.

Terms to Review Before Attempting Worksheet 42

total sum of squares, between groups sum of squares, within groups sum of squares, between groups mean square, within groups mean square, critical value of *F*

Worksheet 42: Computations for a One-Way ANOVA

Riddle: What does "experience" help you recognize?

DIRECTIONS: To find the answer to the riddle, write the answer to each question in the space immediately below it. The word in parentheses in the solution section next to the answer to the first question is the first word in the answer to the riddle, the word beside the answer to the second question is the second word, and so on.

The questions on this worksheet refer to the scores in the following table.

Table 42.6
Scores for Three Groups on a Posttest

Col. 1	Col. 2	Col. 3	Col. 4	Col. 5	Col. 6
Group A	Group B	Group C	A^2	B^2	C^2
3	4	0			
2	5	1			
1	3	1			
5	4	2			
$\sum X_A =$	$\sum X_B =$	$\sum X_C =$	$\sum X_A^2 =$	$\sum X_B^2 =$	$\sum X_C^2 =$
$m =$	$m =$	$m =$			

1. What is the value of the mean for Group A?

2. What is the value of $\sum X_C^2$?

3. What is the value of the within groups sum of squares?

Worksheet 42 (Continued)

4. What is the value of the within groups mean square?

5. What is the observed value of F?

6. Are the differences among the means significant at the .05 level?

7. What decision should be made about the null hypothesis at the .05 level?

Solution section:

2.750 (your) 6.000 (mistakes) 9.084 (ageless) do not reject it (limitless)		
30.917 (royalty) 18.167 (refuses) reject it (again) yes (them)		
4.26 (my) 6.411 (make) 12.750 (when) 1.417 (you) 39 (never)		
66 (interesting) 4.000 (well) 1.000 (younger) 11 (birthday)		

Write the answer to the riddle here, putting one word on each line: _____ _____ _____ _____
_____ _____ _____

Section 43: Tukey's *HSD* Test

As noted in the previous section, when a one-way ANOVA is statistically significant, one cannot be sure which specific differences are significant. Thus, for the example in Section 42, there are three differences between pairs of means:

1. Group A vs. Group B (6.000 − 5.500 = 0.500)
2. Group A vs. Group C (6.000 − 2.333 = 3.667)
3. Group B vs. Group C (5.500 − 2.333 = 3.167)

This *set of differences* was found to be statistically significant at the .05 level.

A number of tests have been suggested for determining the significance of the differences among pairs of means after obtaining an overall significant result using ANOVA.[1] We will examine one of these tests here: ***Tukey's Honestly Significant Difference (HSD) test***. The formula is:

$$HSD = q\sqrt{\frac{MS_{within}}{n}}$$

Where:

q = the studentized range statistic (which we will obtain from Table 8).

MS_{within} is the mean square for within groups calculated for the one-way ANOVA. (See Table 42.5 on page 260, where you will find a value of 2.189 for our example.)

n is the number of individuals in *each* group (in our example from Section 42, there are 6 subjects in each group); thus, $n = 6$. (Note that Tukey's *HSD* test can be used only if we have the same number of individuals in each group.)

---

[1]Because the tests are conducted *after* an ANOVA, they are called *post hoc* tests and are designed for use when individual comparisons were not planned in advance based on theory; they are also known as *multiple comparisons tests*. Although a *t* test may be used to test the difference between a pair of means, its use to make repeated comparisons involving the use of data more than one time makes the probabilities obtained with *t* inaccurate.

263

To use the formula, first determine the value of q as follows. Determine the degrees of freedom associated with the within groups mean square; for our example, it can be read from Table 42.5 on page 260—where we find that it is **15**. Also determine the number of treatments—in our example, this is **3**. Go to Table 8 near the end of this book and read the value of q from the table: Where **3** *treatments* meets a *df* of **15**, we find the value of q for our problem to be **3.67** for the .05 level.[2] Substituting this value of q and the other required values into the formula, we obtain:

$$HSD = 3.67 \sqrt{\frac{2.189}{6}} = \mathbf{2.217}$$

We now compare the value of *HSD* (2.217) to the differences between the pairs of means, which are shown at the beginning of this section. Here is our decision rule:

> If the observed difference between a pair of means is greater than *HSD*, reject the null hypothesis; otherwise, do not reject it.

Here is the result of the comparisons:

1. For Group A vs. Group B, the observed difference is **.500**. Since this is *not* greater than **2.217**, do *not* reject the null hypothesis at the .05 level. The difference is *not* statistically significant at this level.
2. For Group A vs. Group C, the observed difference is **3.667**. Since this is greater than **2.217**, reject the null hypothesis at the .05 level. This difference is statistically significant at this level.
3. For Group B vs. Group C, the observed difference is **3.167**. Since this is greater than **2.217**, reject the null hypothesis at the .05 level. This difference is statistically significant at this level.

---

[2]Use Table 9 for the .01 level.

Thus, the overall one-way ANOVA in Section 42 indicates that there is at least one significant difference in the entire set of differences. In addition, Tukey's *HSD* test indicates that this is because of these significant differences: (1) the difference between Groups A and C and (2) the difference between Groups B and C. The difference between Groups A and B is not significant.

An alternative to Tukey's test is presented in the next section.

Term to Review Before Attempting Worksheet 43

Tukey's Honestly Significant Difference (*HSD*) test

"When our statistics professor isn't in the lab to watch us, he loads up this screen saver."

Worksheet 43: Tukey's *HSD* Test

> *Riddle*: According to Carey Williams, youth is that period when an adolescent knows what?

DIRECTIONS: To find the answer to the riddle, write the answer to each question in the space immediately below it. The word in parentheses in the solution section next to the answer to the first question is the first word in the answer to the riddle, the word beside the answer to the second question is the second word, and so on.

Using the following information, conduct Tukey's HSD tests.

Group A: $m = 2.000, n = 4$
Group B: $m = 2.250, n = 4$
Group C: $m = 6.500, n = 4$

Table 43.1
ANOVA Table

Source of Variation	df	Sum of Squares	Mean Square	F
Between Groups	2	51.167	25.583	12.973*
Within Groups	9	17.750	1.972	
Total	11	68.917		

*$p < .05$

1. What is the size of the observed difference between the means for Groups A and B? (Subtract the smaller from the larger to get a positive value.)

266

Worksheet 43 (Continued)

2. What is the size of the observed difference between the means for Groups A and C? (Subtract the smaller from the larger to get a positive value.)

3. What is the size of the observed difference between the means for Groups B and C? (Subtract the smaller from the larger to get a positive value.)

4. At the .05 level, what is the value of the studentized range statistic (from Table 8) for this problem?

5. What is the value of *HSD*?

6. Is the difference between A and B statistically significant at the .05 level?

7. Is the difference between A and C statistically significant at the .05 level?

8. "The null hypothesis for the difference between B and C may be rejected at the .05 level." Is this statement true or false?

Solution section:

4.500 (everything)	0.250 (almost)	false (school)	3.67 (intelligence)
true (living)	yes (a) no (make)	2.25 (never)	4.250 (but) 9 (helpful)
3.95 (how)	6.50 (something)	2.773 (to)	3.33 (energy) 12.973 (less)

Worksheet 43 (Continued)

Write the answer to the riddle here, putting one word on
each line: _____ _____ _____ _____
_____ _____ _____ _____

Section 44: Scheffé's Test

As noted in the previous two sections, when a one-way ANOVA is statistically significant, one cannot be sure which specific differences are significant. Thus, for the example in Section 42, there are three differences between pairs of means:

1. Group A vs. Group B (6.000 – 5.500 = 0.500)
2. Group A vs. Group C (6.000 – 2.333 = 3.667)
3. Group B vs. Group C (5.500 – 2.333 = 3.167)

This *set of differences* was found to be statistically significant at the .05 level.

A number of tests have been suggested for determining the significance of the differences between pairs of means after obtaining an overall significant result using ANOVA.[1] In Section 43, Tukey's *HSD* test was illustrated. We will examine an alternative test in this section: ***Scheffé's test***. This test is more conservative than Tukey's. That is, *Scheffé's test* is less likely to lead to rejection of the null hypothesis than Tukey's.[2]

To conduct the test, we must apply this formula once for each difference:

$$F = \frac{(m_1 - m_2)^2}{MS_w(n_1 + n_2) \div (n_1)(n_2)}$$

Where:

m_1 is the mean for one of the groups. (To avoid negatives, call the larger of each pair of means m_1.)

m_2 is the mean for another group.

MS_w is the mean square for within groups computed for the one-way ANOVA. (See Table 42.5 on page 260, where the mean square for within groups equals 2.189.)

n_1 is the number of cases in the first group.

---

[1]Because the tests are conducted *after* an ANOVA, they are sometimes called *post hoc tests*; they are also known as *multiple comparison tests*.

[2]There is some disagreement among statisticians as to how to make multiple comparisons; some tests are more conservative than others.

n_2 is the number of cases in the second group.
Application of this formula for Groups A and B is shown below.

For the difference between A and B:

To determine the significance of this difference between the mean of Group A and the mean of Group B, calculate F using the formula shown on page 269, as follows:

$$F_{AB} = \frac{(6.000 - 5.500)^2}{2.189(6+6) \div (6)(6)} = \frac{0.250}{2.189(12) \div 36} = \frac{0.250}{26.268 \div 36} = \frac{0.250}{0.730} = \mathbf{0.342}$$

We now need to compare this observed value of .342 with the critical value. Obtaining the critical value is slightly more complicated that it was for previous tests. First, recall that in Section 42, the degrees of freedom were 2 (for between groups) and 15 (for within groups). Using these, we found in Table 6 (for the .05 level) that the critical value for the one-way ANOVA was 3.68. We use this value in this formula:

$$CV_s = (CV_{ANOVA})(K - 1)$$

Where:

CV_s is the critical value for Scheffé's test.

CV_{ANOVA} is the critical value for the one-way ANOVA.

K is the number of groups (in this case, three: A, B, and C).

Applying it, we find that:

$$CV_s = (3.68)(3 - 1) = (3.68)(2) = \mathbf{7.36}$$

Our decision rule is:

If the *observed value is greater than the critical value,*[3] reject the

---

[3]In the unlikely event that the observed value is exactly equal to the critical value, reject the null hypothesis.

Because our observed value (**.342**) is *not* greater than the critical value (**7.36**), we do *not* reject the null hypothesis at the .05 level; the difference between the means of Groups A and B is *not* statistically significant.

For the difference between A and C:

$$F_{AC} = \frac{(6.000 - 2.333)^2}{2.189(6+6) \div (6)(6)} = \textbf{18.421}$$

To evaluate this value of *F*, we use the same critical value that we calculated above (**7.36**) and the same decision rule. Because the value of F_{AC} (**18.421**) exceeds the critical value (**7.36**), we reject the null hypothesis at the .05 level; the difference between the means of Groups A and C is statistically significant.

For the difference between B and C:

$$F_{BC} = \frac{(5.500 - 2.333)^2}{2.189(6+6) \div (6)(6)} = \textbf{13.740}$$

To evaluate this value of *F*, we use the same critical value that we calculated above (**7.36**) and the same decision rule. Because the value of F_{BC} (**13.740**) exceeds the critical value (**7.36**), we reject the null hypothesis at the .05 level; the difference between the means of Groups B and C is statistically significant.

Thus, we have arrived at the same conclusions regarding the significance of the difference between the pairs of means that we arrived at using Tukey's *HSD* test in Section 43. The overall one-way ANOVA in Section 42 indicated that the entire set of differences is statistically significant. Scheffé's test has indicated that this was the result of the differences between (1) Groups A and C and (2) Groups B and C. We should not discuss the difference between Groups A and B as being significant.

Although the tests in Sections 43 and 44 led to the same conclusion in our example, they will not always do so. There may be times when Tukey's *HSD* results in the rejection of a null hypothesis for the difference between a pair of means when Scheffé's test does not. Use of either test is acceptable; however, a consumer of your research may feel more comfortable with your rejections of the null hypothesis if you use the more conservative Scheffé's test. Another consideration in selecting between the two tests is that you do not have to have equal numbers of individuals to use Scheffé's test, which you must have to use Tukey's *HSD*.

Term to Review Before Attempting Worksheet 44

Scheffé's test

Worksheet 44: Scheffé's Test

> *Riddle*: According to Edward Gibbon, beauty is seldom despised except by whom?

DIRECTIONS: To find the answer to the riddle, write the answer to each question in the space immediately below it. The word in parentheses in the solution section next to the answer to the first question is the first word in the answer to the riddle, the word beside the answer to the second question is the second word, and so on.

Use the data at the beginning of Worksheet 43. Conduct Scheffé's test at the .05 level.

1. What is the value of F for the difference between the means of Groups A and B?

2. What is the critical value for Scheffé's test at the .05 level?

3. Is the difference between the means of Groups A and B significant at the .05 level?

4. What is the value of F for the difference between the means of Groups A and C?

5. May the null hypothesis be rejected for the difference between the means of Groups A and C?

Worksheet 44 (Continued)

6. What is the value of F for the difference between the means of Groups B and C?

7. "The difference between the means for Groups B and C is statistically significant." Is this statement true or false?

Solution section:

18.319 (been)	yes (has)	20.538 (it)	false (sudden)	0.986 (contest)
4.26 (willing)	true (refused)	4.564 (face)	8.52 (to)	0.064 (those)
no (whom)	0.255 (never)	15.333 (asking)	0.006 (truthfully)

Write the answer to the riddle here, putting one word on each line: _____ _____ _____ _____ _____ _____ _____

Section 45: Introduction to Two-Way ANOVA

In a *two-way ANOVA* (also known as a *two-factor ANOVA*) subjects are classified in two ways. Consider Example 1, which illustrates a two-way ANOVA.

Example 1:

A random sample of welfare recipients was assigned to a new job training program. Another random sample was assigned to a conventional job training program. (*Note*: Which of the programs they were assigned to is one of the ways in which the subjects were classified.) Subjects were also classified according to whether or not they had a high school diploma. All of the subjects in each group found employment in the private sector at the end of their training. Their mean hourly wages are shown in this table:[1]

	Type of Program		
	Conventional	New	**Row Means**
H.S. Diploma	$m = \$8.88$	$m = \$8.75$	$m = \$8.82$
No H.S. Diploma	$m = \$4.56$	$m = \$8.80$	$m = \$6.68$
Column Means	$m = \$6.72$	$m = \$8.78$	

First, let's consider the column means of $6.72 (for the conventional program) and $8.78 (for the new program). These suggest that, overall, the new program is superior to the conventional one. In other words, if we temporarily ignore whether subjects have a high school diploma, the new program seems

---

[1]Note that income in large populations is usually skewed, making the mean an inappropriate average (see Section 11); for these groups, assume that it was not skewed. Also note that the row means and column means were obtained by adding and dividing by two; this is appropriate only if the number of subjects in all cells is equal; if it is not, compute the row and column means using the original raw scores.

superior to the conventional one. This difference ($8.78 – $6.72 = $2.06) suggests that there is what is called a **main effect**. A *main effect* is the result of comparing one of the ways in which the subjects were classified while temporarily ignoring the other way in which they were classified.

Since the concept of *main effect* is important, let's look at it another way. You can see that the column mean of $6.72 is for all those who had the conventional program regardless of whether they have a high school diploma. The column mean of $8.78 is for all those who had the new program regardless of whether they have a high school diploma. Thus, by looking at the column means, we are considering only the effects of the type of program (and *not* the effects of a high school diploma). When you look at the effects of only one way in which the subjects were classified, this is called a *main effect*.

Now consider the row means of $8.82 (for those with a high school diploma) and $6.68 (for those with no high school diploma). This suggests that those with a diploma, on the average, have higher earnings than those without one. This is also a *main effect*. This main effect is for the "diploma vs. no diploma variable"—while temporarily ignoring the type of training program.

To this point, we have two findings that would be of interest to those studying welfare: (1) The new program seems to be superior to the conventional program in terms of hourly wages and (2) those with a high school diploma seem to have higher hourly wages. (*Note*: The term "seem" is being used because we have only studied random samples and we do not yet know whether the differences are statistically significant.)

You may have already noticed that there is a third interesting finding: Those with a high school diploma earn about the same amount regardless of the program. This statement is based on these means for *those with a high school diploma* reproduced from the table shown above:

	Conventional Program	**New Program**
H.S. Diploma	$m = \$8.88$	$m = \$8.75$

However, those with no high school diploma seem to benefit more from the new program than the conventional one. This statement is based on these means for *those with no high school diploma*:

	Conventional Program	New Program
No H.S. Diploma	$m = \$4.56$	$m = \$8.80$

Now suppose you were the researcher who conducted this study; you are now an expert on the subject and an administrator calls you for advice. She asks you, "Which program should we use? The conventional one or the new one?" You could, of course, tell her that there is a *main effect* for programs that suggests that, overall, the new program is superior in terms of wages—but, if you stopped there, your answer would be incomplete. A more complete answer is:

1. For those with a diploma, the two programs are about equal in effectiveness. Thus, the choice of a program for them should probably hinge on other considerations such as the cost of the two programs.
2. For those with no diploma, the new program is superior to the conventional one. Other things being equal, those without a diploma should be assigned to the new program.

Because you cannot give a complete answer about the two types of programs (one way in which the subjects were classified) without also referring to high school diplomas (the other way in which they were classified), we say there is an ***interaction*** between the two. How well the two programs work depends, in part, upon whether the subjects have high school diplomas.

Here is a simple way in which you can spot an interaction when there are only two rows of means: Subtract each mean in the second row from the mean in the first row. If the two differences are the same, there is no interaction. If they are different, there is an interaction. Here is how it works for the data in Example 1:

	Type of Program	
	Conventional	New
H.S. Diploma	$m = \$8.88$	$m = \$8.75$
No H.S. Diploma	$m = \$4.56$	$m = \$8.80$
Difference	**$4.32**	**−$0.05**

Because the two differences are *not* the same, there is an interaction.

Consider Example 2, in which there are no main effects but there is an interaction.

Example 2:

A random sample of subjects from a population of those suffering from a chronic illness was administered a new drug. Another random sample from the same population was administered a standard drug. Subjects were also classified as to whether they were male or female. At the end of the study, improvement was measured on a scale from 0 (for no improvement) to 10 (for complete recovery). These means were obtained:

	Drug		Row Means
	Standard	New	
Male	$m = 5.00$	$m = 7.00$	**m = 6.00**
Female	$m = 7.00$	$m = 5.00$	**m = 6.00**
Column Means	**m = 6.00**	**m = 6.00**	

The two column means in Example 2 are the same. Thus, if we temporarily ignore whether subjects are male or female, we would conclude that the two drugs are equally effective. To state it statistically, we would say that *there is no main effect for the drugs.*

The two row means are also the same. Thus, if we temporarily ignore which drug was taken, we can conclude that males and females improved to the same extent. To state it statistically, we would say that *there is no main effect for gender*.

Of course, the interesting finding in Example 2 is the *interaction*. The standard drug works better for females and the new drug works better for males. Subtracting as we did in Example 1, we obtain the differences shown below. Because –2.00 is not equal to 2.00, there is an interaction.

	Drug	
	Standard	New
Male	$m = 5.00$	$m = 7.00$
Female	$m = 7.00$	$m = 5.00$
Difference	**–2.00**	**2.00**

Consider Example 3, in which there are two main effects but no interaction.

Example 3:

Random samples of high and low achievers were assigned to one of two types of reinforcement during math lessons. Achievement on a math test at the end of the experiment was the outcome variable. The mean scores on the test were:

	Type of Reinforcement		
	Type A	Type B	**Row Means**
High achievers	$m = 50.00$	$m = 30.00$	$m = 40.00$
Low achievers	$m = 40.00$	$m = 20.00$	$m = 30.00$
Column Means	$m = 45.00$	$m = 25.00$	

In Example 3, there seems to be a main effect for type of reinforcement as indicated by the difference between the column means (45.00 and 25.00). Thus, ignoring achievement levels temporarily, Type A seems to be more effective than Type B.

There also seems to be a main effect for achievement level as indicated by the difference between the row means (40.00 and 30.00). Thus, ignoring the type of reinforcement, high achievers score higher on the math test than low achievers.

There is no interaction, as indicated by the differences, which are:

| | Type of Reinforcement | |
	Type A	Type B
High achievers	$m = 50.00$	$m = 30.00$
Low achievers	$m = 40.00$	$m = 20.00$
Difference	**10.00**	**10.00**

What does this lack of an interaction tell us? That regardless of the type of reinforcement, high achievers are the same number of points higher than low achievers (i.e., 10 points). Put another way, regardless of whether students are high or low achievers, Type A reinforcement is better.[2]

We are not restricted to two categories for each classification variable. We could, for example, study three types of reinforcement and obtain means for the following cells and still consider the two main effects and the interaction.

| | Type of Reinforcement | | | |
	Type A	Type B	Type C	**Row Means**
High	$m =$	$m =$	$m =$	$m =$
Low	$m =$	$m =$	$m =$	$m =$
Column Means	$m =$	$m =$	$m =$	

---

[2]The basis for this second statement is that if you subtract across the rows, you get the same difference for each row. Earlier, you were told to subtract down columns; however, subtracting across the rows works equally well in determining whether there is an interaction.

In review, a two-way ANOVA examines two *main effects* and one *interaction*. Of course, because only random samples have been examined, the null hypothesis must be considered. For each of the main effects and for the interaction, the null hypothesis states that there is no *true* difference—that the observed differences were created by random sampling errors. A two-way ANOVA will, therefore, test the two main effects and the interaction for significance. This is done by computing three values of F (one for each of the three null hypotheses) and determining the probability associated with each. Typically, if a probability is .05 or less, the null hypothesis is rejected and the main effect or interaction is declared to be statistically significant.

Although the computational procedures for a two-way ANOVA are beyond the scope of this book, all standard statistical software packages permit such an analysis. Because computations for a two-way ANOVA are complex, use of statistical software is recommended.

Terms to Review Before Attempting Worksheet 45

two-way ANOVA (two-factor ANOVA), main effect, interaction

Worksheet 45: Introduction to Two-Way ANOVA

> *Riddle*: A banker is someone who will lend you an umbrella when the sun is out but...

DIRECTIONS: To find the answer to the riddle, write the answer to each question in the space immediately below it. The word in parentheses in the solution section next to the answer to the first question is the first word in the answer to the riddle, the word beside the answer to the second question is the second word, and so on.

Assume there are equal numbers of individuals in each cell in all examples.

Questions 1 through 3 refer to this information.
Random samples of those with headache pain and those with muscular pain were randomly assigned to two types of pain relievers. The means indicate the average amount of pain relief for each condition.

1. Does there seem to be a main effect for type of pain reliever?

	Type of Pain Reliever		
	Type A	Type B	**Row Means**
Headache pain	$m = 20.00$	$m = 30.00$	$m =$
Muscular pain	$m = 15.00$	$m = 25.00$	$m =$
Column Means	$m =$	$m =$	

2. Does there seem to be a main effect for type of pain (headache vs. muscular)?
   (Circle one letter.)

   A. yes

   B. no

Worksheet 45 (Continued)

3. Does there seem to be an interaction?

4. If a two-way ANOVA was conducted, at what probability level would a null hypothesis typically be rejected (i.e., what is the lowest level at which significance is routinely declared)?

Questions 5 through 7 refer to this information.

Two types of piano instruction were tried with random samples of individuals who either had previous instruction on playing or did not have previous instruction. The means indicate the proficiency at playing the piano at the end of the treatments.

	Type of Piano Instruction		
	Type A	Type B	**Row Means**
Had previous instruction	$m = 130.0$	$m = 100.0$	$m =$
Had no previous instruction	$m = 100.0$	$m = 130.0$	$m =$
Column Means	$m =$	$m =$	

5. "There seem to be no main effects." Is this statement true or false?

6. "There seems to be no interaction." Is this statement true or false?

Worksheet 45 (Continued)

7. Which of the following *F* tests would most likely be significant? (Circle one letter.)

    A. The test for the type of piano instruction.

    B. The test for previous vs. no previous instruction.

    C. The test for the interaction.

Questions 8 and 9 refer to this information.

Random samples of rats were assigned to either a food reward or no reward. In addition, they were randomly assigned to either run a maze in the dark or in daylight. The means indicate the average number of seconds it took each group to run a standard maze.

	Type of Reward		
	Food	No Reward	**Row Means**
Dark	$m = 2.4$	$m = 3.4$	$m =$
Daylight	$m = 1.4$	$m = 2.4$	$m =$
Column Means	$m =$	$m =$	

8. There seems to be a main effect for (Circle one letter.)

    D. food vs. no reward.

    E. dark vs. daylight.

    F. both D and E.

9. There seems to be (Circle one letter.)

    G. an interaction.

    H. no interaction.

Worksheet 45 (Continued)

Solution section:

yes (wants)   no (back)   A. (it)   2 (money)   H. (rain)   F. (to)

B. (lender)   C. (begins)   false (it)   true (minute)   .05 (the)

1 (being)   G. (checks)   D. (statement)   E. (sometimes)

Write the answer to the riddle here, putting one word on each line: _____ _____ _____ _____

_____ _____ _____ _____

"This is my relaxation tape——it's the sound of ocean waves crashing onto the shore, snatching my statistics book off my beach chair, and carrying it out to sea."

Notes:

Section 46: Significance of the Difference Between Variances

Sections 35 through 45 deal with testing the differences among means for significance. Occasionally, we are interested in the differences between variances. Here is an example:

A randomly selected sample of students was assigned to learn elementary algebra using a new computer-assisted program (CAP); this was the experimental group. Another randomly selected sample was designated as the control group; it received elementary algebra instruction via a traditional textbook approach. The researcher hypothesized that the posttest scores of the experimental group would be more variable than those of the control group. The rationale for this research hypothesis was that the CAP allows students to move at their own pace—so that high-achieving students could zip ahead while low-achieving students could move slowly. In contrast, the traditional approach would hold back the high achievers and pull up the low achievers, leading to less variability in the control group's scores. The following statistics were obtained:

Experimental	Control
$m = 43.22$	$m = 42.99$
$s = 7.88$	$s = 5.22$
$n = 21$	$n = 20$

You should recall that the standard deviation is the most widely used measure of variability. Comparison of the two standard deviations reveals an observed difference that is consistent with the research hypothesis—the standard deviation of the experimental group ($s = 7.88$) is greater than that of the control group ($s = 5.22$). However, before reporting the results, the researcher should consider the null hypothesis, which says that the observed difference was created by random sampling errors.[1]

To test the null hypothesis, first compute F using this formula:

$$F = \frac{S_L^2}{S_S^2}$$

Where:

S_L^2 is the larger standard deviation squared.

S_S^2 is the smaller standard deviation squared.

You might recall that the squared standard deviation has its own name: the **variance**. Hence, strictly speaking, we are testing the difference between the variances. Of course, the size of the variance is directly related to the size of the standard deviation.

Applying the formula to our example, we obtain:

$$F = \frac{7.88^2}{5.22^2} = \frac{62.09}{27.25} = \mathbf{2.28}$$

Thus, our **observed value** of F is **2.28**. We need to compare it with the **critical value** of F to determine significance. To obtain the critical value, we must first compute degrees of freedom. For the group with the larger variance:

$$df = n - 1 = 21 - 1 = \mathbf{20}$$

Thus, **20** is the degrees of freedom associated with the **numerator** of the formula for F.

For the group with the smaller variance:

$$df = n - 1 = 20 - 1 = \mathbf{19}$$

---

[1]Note that in the term "random sampling errors" the adjective "random" is superfluous because statisticians define "sampling error" as "error created by random sampling." The redundant term is used here for instructional purposes.

Thus, **19** is the degrees of freedom associated with the **denominator** of the formula for F.

For the .05 level,[2] we go to Table 6 to get the critical value of F. Notice that the columns are labeled "Between Groups Degrees of Freedom (**Numerator**)." The word *numerator* is included to help you with this significance test. Look up the *column* for the degrees of freedom associated with the numerator, which is **20**. Then look up the *row* for the degrees of freedom for the denominator, which is **19**. Where *column 20* intersects with *row 19*, we find a critical value of **2.15**.

Here is our decision rule:

If the *observed value is greater than the critical value,*[3] reject the null hypothesis; otherwise, do not reject it.

Because our observed value (**2.28**) is greater than the critical value (**2.15**), we reject the null hypothesis and declare the difference between the two variances to be statistically significant at the .05 level.[4] Thus, we have ruled out the null hypothesis as an explanation for the difference between the variance of the experimental group and the variance of the control group.

Terms to Review Before Attempting Worksheet 46

variance, observed value, critical value

---

[2] Use Table 7 for the .01 level.

[3] In the unlikely event that the observed value is exactly equal to the critical value, reject the null hypothesis.

[4] You may have noticed that some of the possible degrees of freedom are missing in Table 6. For example, rows 66 through 69 are missing. This is because there are only very small differences in the critical values among these values. Suppose you had 67 degrees of freedom. Use the smaller of the ones shown—in this case, 66, which is conservative, meaning that you have very slightly reduced the odds of rejecting the null hypothesis.

Worksheet 46: Significance of the Difference Between Variances

> *Riddle*: What's the loudest sound that you will ever hear?

DIRECTIONS: To find the answer to the riddle, write the answer to each question in the space immediately below it. The word in parentheses in the solution section next to the answer to the first question is the first word in the answer to the riddle, the word beside the answer to the second question is the second word, and so on.

Questions 1 through 5 refer to this information.

Experimental	Control
$m = 25.44$	$m = 26.99$
$s = 6.77$	$s = 5.89$
$n = 17$	$n = 15$

1. Which group ("experimental" *or* "control") has more variance?

2. Which standard deviation ("6.77" *or* "5.89") should you enter into the numerator of the formula for F?

3. To two decimal places, what is the observed value of F?

4. At the .05 level, what is the critical value of F for this problem?

Worksheet 46 (Continued)

5. Should the null hypothesis be rejected at the .05 level?

Questions 6 through 8 refer to this information.

Experimental	Control
$m = 100.48$	$m = 102.65$
$s = 10.03$	$s = 15.49$
$n = 30$	$n = 31$

6. To two decimal places, what is the observed value of F?

7. At the .05 level, what is the critical value of F for this problem?

8. Is the difference between the variances significant at the .05 level?

Solution section:

yes (car)   control (breaking)   .76 (hearing)   experimental (the)

6.77 (first)   no (your)   2.39 (brand)   5.89 (shattering)   .41 (screams)

1.85 (new)   1.32 (rattle)   2.44 (in)   8.00 (listening)   9.00 (record)

10.00 (office)   7.89 (yelling)   2.00 (ears)   .51 (everything)

Write the answer to the riddle here, putting one word on each line: _____ _____ _____ _____ _____ _____ _____ _____

**Statistics Laboratory
Hours 9AM - 9PM**

GLASBERGEN

**"And this button gives the computer a
mild electric shock when I need to punish it."**

Section 47: Significance of a Pearson *r*

Often, we want to know whether a Pearson *r* is significant. This question arises when we have drawn a random sample from a population, measured two variables, and have determined the correlation between the two sets of scores using a Pearson *r*. Here is an example:

Twenty students were drawn at random from a population and their GPAs were correlated with their heights. A Pearson *r* of .23 was obtained, indicating that those with higher GPAs are taller.

Before reporting this result, the researcher needs to take into account that her ***observed value*** of *r* (.23) is based on a random sample of only 20 students; therefore, the result may not be true in the population. Thus, she needs to test the null hypothesis.[1]

The test of the null hypothesis for this type of problem is a special version of the *t* test. Fortunately, we do not actually have to compute *t*; instead, we can take a shortcut and determine the significance of a given correlation coefficient by referring to the values in Table 10 near the end of this book.[2]

To use Table 10, first compute the degrees of freedom, using this formula:

$$df = n - 2$$

Where:

n is the number of subjects.

Since we have 20 subjects in our example, $df = 20 - 2 = \mathbf{18}$.

We look up **18** degrees of freedom in Table 10 and find a ***minimum value*** of *r* for significance at the .05 level of **.444**.

---

[1]In this case, we are concerned with the possibility that in the population, there is a true correlation of 0.00. The null hypothesis for this situation says that the difference between 0.00 (the hypothesized true correlation) and 0.23 (the observed correlation) was caused by sampling errors as a result of random selection.

[2]The values in Table 10 were determined using the special *t* test.

Here is our decision rule:

If the *observed value of r is greater than the minimum value*[3] in Table 10, reject the null hypothesis; otherwise, do not reject it.[4]

Since the observed value of **.23** is *not* greater than the minimum value of **.444**, we do *not* reject the null hypothesis; the difference is *not* statistically significant at the .05 level. Thus, the researcher has failed to demonstrate a significant correlation between GPA and height.

Terms to Review Before Attempting Worksheet 47

observed value, minimum value

"You can live a perfectly normal life if you accept the fact that statistics prove that your life will never be perfectly normal."

---

[3]In the unlikely event that the observed value is exactly equal to the minimum value, reject the null hypothesis.

[4]Notice that in Table 10 only selected values of *df* are given. If you obtain a *df* that is not shown in the table, use the next lower value shown. For example, if you have 52 degrees of freedom (which is not shown) use 50 (which is shown).

Worksheet 47: Significance of a Pearson *r*

> *Riddle*: What's the trouble with American Thanksgiving dinners?

DIRECTIONS: To find the answer to the riddle, write the answer to each question in the space immediately below it. The word in parentheses in the solution section next to the answer to the first question is the first word in the answer to the riddle, the word beside the answer to the second question is the second word, and so on.

1. For a random sample of 52 subjects, a Pearson *r* of .29 was obtained for the relationship between variables A and B. What is the value of the degrees of freedom for determining the significance of this *r*?

2. What is the minimum value of *r* for significance at the .05 level for the study described in Question 1?

3. Is the value of *r* in Question 1 statistically significant at the .05 level?

4. Is the value of *r* in Question 1 statistically significant at the .01 level?

5. For a random sample of 14 subjects, a Pearson *r* of .29 was obtained for the relationship between variables X and Y. What is the value of the degrees of freedom for determining the significance of this *r*?

Worksheet 47 (Continued)

6. "The value of the Pearson *r* in Question 5 is statistically significant at the .05 level." Is this statement true or false?

7. For a random sample of 47 subjects, a Pearson *r* of .39 was obtained for the relationship between variables X and Y. "This *r* of .39 is statistically significant at the .01 level." Is this statement true or false?

Solution section:

50 (two) 51 (eating) 52 (turkey) true (again) false (hungry)

14 (stuffed) .273 (days) yes (later) 13 (family) no (you)

.354 (sleepy) .29 (cooking) 12 (are) .39 (sensation) .01 (nervous)

Write the answer to the riddle here, putting one word on each line: _____ _____ _____ _____
_____ _____ _____

Section 48: Introduction to Chi Square

Frequently, our research data is nominal (i.e., naming data such as subjects naming the political candidates for whom they plan to vote).[1] Such data do not directly permit the computation of means and standard deviations. Instead, we usually report the number of subjects who named each category (i.e., the frequency) and the corresponding proportions or percentages. Here is an example:

Example 1:
A random sample of 200 registered voters was drawn and asked which of two candidates for an elected office they planned to vote. These data were obtained:

<u>Candidate Smith</u>   <u>Candidate Doe</u>
$n = 110$ (55.0%) $n = 90$ (45.0%)

The data in Example 1 suggest that Candidate Smith is leading. However, only a random sample of voters was surveyed. It is possible, for example, that the population of voters is evenly split, but that a difference of 10 percentage points was obtained because of the sampling errors associated with random sampling. For this possibility, there is a null hypothesis that says that there is no true difference in the population; that is, in the population, the voters are evenly split. We cannot use a t test or ANOVA to test this null hypothesis because we do not have means. The appropriate test for the data under consideration (i.e., frequencies or numbers of cases) is ***chi square***,[2] whose symbol is χ^2. As it turns out for these data, the probability that the null hypothesis is true is greater than 5 in 100 ($p > .05$).[3] Thus, we cannot reject the null hypothesis; the

---

[1] See Section 2 to review scales of measurement, including nominal.
[2] The tests on means (t and F) in earlier sections are based on the assumption that the underlying distributions are normal; these are examples of *parametric tests*. Since chi square is not based on such an assumption, it is an example of a *nonparametric* (or *distribution-free*) test.
[3] The procedures for calculating χ^2 and obtaining the probabilities are described in the next two sections.

difference is not statistically significant. In concrete terms, Candidate Smith cannot rest easily because we cannot rule out sampling errors as an explanation for the difference in her favor.

Example 1 illustrates a ***one-way chi square*** (also known as a ***goodness of fit chi square***). The subjects are classified in only one way—for whom they plan to vote. Example 2 illustrates a ***two-way chi square*** in which samples from two populations of voters (that is, males and females) were classified in terms of for whom they plan to vote.[4]

Example 2:

A random sample of 200 male registered voters and a random sample of 200 female registered voters were drawn and asked for which of two candidates for an elected office they plan to vote. These data were obtained:

	Candidate Jones	Candidate Black
Males	$n = 80$	$n = 120$
Females	$n = 120$	$n = 80$

Inspection of the data in Example 2 suggests that Jones is a stronger candidate among females and Black is a stronger candidate among males. If this pattern is true among all males and all females in the voting population, both candidates should take heed. For example, Candidate Jones might consider ways to shore up her support among males without alienating the females; Candidate Black might do the opposite. However, only a random sample was surveyed. Before taking action, the candidates should consider how likely it is that the observed differences in preferences between the two groups (males and females) were created by random sampling errors. Chi square is the appropriate test because we are examining differences in nominal data. For this example,

[4]There are two types of two-way chi square tests. Example 2 illustrates a *chi square test of homogeneity*. This test involves two or more populations (e.g., males and females) and their opinions on one outcome variable (e.g., for which candidate they plan to vote). Example 3 illustrates a *chi square test of independence* in which one population is classified in two ways.

chi square reveals that p is less than 1 in 1,000 ($p < .001$). Thus, the probability that random errors created the differences is less than 1 in 1,000. In other words, it is very unlikely that this pattern of differences is due to sampling errors—with a high degree of confidence, the candidates can rule out chance as an explanation.[5]

In Example 3, a sample from one population of subjects was asked two questions, each of which yielded nominal data.

Example 3:

A random sample of college students was asked whether they think that *IQ* tests measure innate intelligence and whether they had taken a tests and measurements course. These data resulted:

	Took Course	Did Not Take Course
Yes, Innate	$n = 20$	$n = 30$
No, Not Innate	$n = 40$	$n = 15$

The observed data suggest that those who did not take the course were more likely to perceive *IQ* tests as measuring innate intelligence (30 vs. 15) than those who took the course (20 vs. 40). In other words, there appears to be a relationship between whether subjects have taken the course and what they believe *IQ* tests measure. Once again, only a random sample was questioned and, thus, it is possible that the observed relationship is not true in the population. That is, the null hypothesis asserts that there is no *true* relationship (in the population). A chi square test for these data produce this result:

$$\chi^2 = 11.455,\ df = 1,\ p < .001$$

---

[5]Notice that in our examples, the responses are independent. For instance, in Example 2, the gender of a person is not determined by their preference for a candidate. Also, each response is mutually exclusive. For example, we do not allow a subject to indicate that he/she is both male and female. Independence of mutually exclusive categories are assumptions underlying chi square.

Thus, the null hypothesis may be rejected with a high degree of confidence since the likelihood is less than 1 in 1,000 that it is a true hypothesis.

Section 49 illustrates the computational procedures for conducting a one-way chi square, and Section 50 illustrates the procedures for a two-way chi square.

Terms to Review Before Attempting Worksheet 48

chi square, χ^2, one-way chi square (goodness of fit chi square),
two-way chi square,

"Whenever I get a wrong answer to a statistics problem,
I just push this little button and restart.
I wish my whole life was like that!"

Worksheet 48: Introduction to Chi Square

> *Riddle*: What advice did Erma Bombeck offer about selecting a doctor?

DIRECTIONS: To find the answer to the riddle, write the answer to each question in the space immediately below it. The word in parentheses in the solution section next to the answer to the first question is the first word in the answer to the riddle, the word beside the answer to the second question is the second word, and so on.

1. If you calculated the mean score for a sample of boys and the mean score for a sample of girls on a standardized test and wanted to compare the two means with an inferential test, would a chi square test be appropriate?

2. If you asked a random sample of subjects which of two brands of coffee they preferred and wanted to compare the frequencies with an inferential test, would a chi square test be appropriate?

3. Suppose subjects were asked whether they had gotten a flu shot before the beginning of the flu season and then asked at the end of the flu season whether they had gotten the flu in order to examine the relationship between getting a shot and getting the flu. Is this a "one-way chi square" *or* a "two-way chi square" problem?

4. Suppose you read that $\chi^2 = 4.001$, $df = 1$, $p < .05$. What decision should be made about the null hypothesis at the .05 level?

Worksheet 48 (Continued)

5. "If $\chi^2 = 10.999$, $df = 2$, $p < .01$, then the differences are statistically significant at the .01 level." Is this statement true or false?

6. "If $\chi^2 = 2.578$, $df = 1$, $p > .05$, then the differences are statistically significant at the .05 level." Is this statement true or false?

7. Suppose that as a result of a chi square test, p is found to be less than .001 for a given set of data. This means that the likelihood that the null hypothesis is correct is less than 1 in ___?

8. If a difference is found to be statistically insignificant with chi square, what decision should be made about the null hypothesis?

Solution section:

100 (wine)   10,000 (sleeping)   one-way (nevertheless)
do not reject it (died)   1,000 (have)   false (plants)   true (office)
reject it (whose)   two-way (doctor)   no (avoid)   yes (a)

Write the answer to the riddle here, putting one word on each line: _____ _____ _____ _____
_____ _____ _____ _____

Section 49: Computations for a One-Way Chi Square

As noted in the previous section, a one-way chi square is one in which a random sample of subjects is classified on a single variable. For example, we might have three world history textbooks that are being considered for use in the high schools of a state. A random sample of the history teachers is asked to examine the three books and to answer a number of questions about them. The crucial question is, overall, which book do they prefer? These are their preferences:

Textbook A	Textbook B	Textbook C
$n = 30$ (37.97%)	$n = 27$ (34.18%)	$n = 22$ (27.85%)

The results suggest that Textbook A is preferred to Textbooks B and C and that Textbook B is preferred to Textbook C. But the researchers have examined only a random sample of all history teachers. Thus, it is possible that if all the teachers had examined the books, there would have been no difference in preference for the three books. This possibility leads to the null hypothesis that there is no *true* difference in preference (in the population).[1] We can test this null hypothesis with a chi square test.

The frequencies that we obtained from our sample are called the ***observed frequencies*** (i.e., the frequencies that we actually observed in a study—in this case, 30, 27, and 22). The symbol for the observed frequencies is ***O***. The frequencies that are expected based on the null hypothesis are called the ***expected frequencies*** (these are the frequencies we should expect to obtain if the null hypothesis is correct). The symbol for the expected frequencies is ***E***. Because

---

[1]In this example, we are going to test the null hypothesis that the population of teachers is evenly split. A one-way chi square may also be used to test the difference between observed frequencies and the frequencies that might be expected based on a theory or information at hand about a population. For example, if we know that 60% of the voters in a state are registered as Democrats and 40% are registered as Republicans, and if we draw a random sample of 100 voters from one county in the state and obtain 55% Democrats and 45% Republicans, this question arises: Are the voters in the county significantly different from the voters statewide? The null hypothesis for this problem states that there is no true difference between the differences in registration in the sample and the differences in registration in the population.

the null hypothesis for the study under consideration states that there is no difference in preference, we would expect equal frequencies for the three textbooks. Since there are 79 subjects in the example (30 + 27 + 22 = 79), we would expect 26.333 subjects to prefer each textbook (i.e., 79 subjects divided by 3 textbooks = 26.333).[2]

Let us review the steps up to this point for solving a one-way chi square problem:

1. Select a random sample and classify subjects in a single way such as which textbook they prefer. You may have as many categories as needed for the problem—in our example, there are three categories. Call the frequencies (i.e., number of cases) in each category the *observed frequencies* (*O*).

2. Calculate the *expected frequencies* (*E*) for the null hypothesis under consideration. If your null hypothesis states that there are no differences in the population, divide the total number of subjects in the sample by the number of categories—in our example, there are 79 subjects and 3 categories; thus, 79/3 = 26.333, which is the expected frequency for each category. These data are shown in Table 49.1.

Table 49.1
Observed and Expected Frequencies

	Textbook A	Textbook B	Textbook C
Observed	30	27	22
Expected	26.333	26.333	26.333

The next step is to apply the formula for chi square:

$$\chi^2 = \Sigma \frac{(O-E)^2}{E}$$

[2]For the example in the first footnote, the expected frequencies would be 60 Democrats (60% of the 100 subjects in the sample) and 40 Republicans (40% of the 100 subjects in the sample).

For Textbook A: $\dfrac{(30-26.333)^2}{26.333} = \dfrac{(3.667)^2}{26.333} = \dfrac{13.447}{26.333} = \mathbf{0.511}$

For Textbook B: $\dfrac{(27-26.333)^2}{26.333} = \dfrac{(0.667)^2}{26.333} = \dfrac{0.445}{26.333} = \mathbf{0.017}$

For Textbook C: $\dfrac{(22-26.333)^2}{26.333} = \dfrac{(-4.333)^2}{26.333} = \dfrac{18.775}{26.333} = \mathbf{0.713}$

Summing the values for the three textbooks yields the value of chi square:

$$\chi^2 = 0.511 + 0.017 + 0.713 = \mathbf{1.241}$$

Thus, **1.241** is our *observed value of chi square*.

Before we can determine significance, we need to compute the degrees of freedom. For a one-way chi square the formula is:

df = number of categories − 1

Because we have three categories of textbooks, the degrees of freedom for our example are:

$df = 3 - 1 = \mathbf{2}$

To determine significance, compare the ***observed value of chi square*** that we computed with the ***critical value of chi square*** in Table 11. In this table, we find that for 2 degrees of freedom, the critical value at the .05 level is **5.991**. Here is the decision rule:

> If the *observed value of chi square is greater than the critical value,*[3] reject the null hypothesis; otherwise, do not reject it.

---

[3]In the unlikely event that the observed value is exactly equal to the critical value, reject the null hypothesis.

Because our observed value (1.241) is *not* greater than the critical value (5.991), we do *not* reject the null hypothesis at the .05 level. We must conclude that the differences are *not* statistically significant.

Terms to Review Before Attempting Worksheet 49

**observed frequencies (*O*), expected frequencies (*E*),
observed value of chi square, critical value of chi square**

**"I'm going to use hypnosis to help you confront repressed
traumatic memories of your last statistics class."**

Worksheet 49: Computations for a One-Way Chi Square

Riddle: How do you calculate how much education you have?

DIRECTIONS: To find the answer to the riddle, write the answer to each question in the space immediately below it. The word in parentheses in the solution section next to the answer to the first question is the first word in the answer to the riddle, the word beside the answer to the second question is the second word, and so on.

A random sample of 99 social workers was asked to indicate which of three welfare proposals they preferred. The following data were observed:

Proposal A	Proposal B	Proposal C
29	31	39

Compute chi square and test the null hypothesis that there is no true difference in preferences in the population.

1. What is the expected frequency for Proposal A?

2. To three decimal places, what is the observed value of chi square? (Allow for minor variations in the third place due to rounding.)

3. What is the value of the degrees of freedom?

4. At the .05 level, what is the critical value of chi square?

Worksheet 49 (Continued)

5. Is chi square significant at the .05 level?

6. What decision should be made about the null hypothesis at the .05 level?

Questions 7, 8, and 9 refer to the following information.

These results were obtained for another research problem:
$\chi^2 = 10.22$, $df = 3$.

7. "Chi square is significant at the .05 level." Is this statement true or false?

8. What decision should be made about the null hypothesis at the .05 level?

9. "Chi square is significant at the .01 level." Is this statement true or false?

Solution section:

yes (college) 1.091 (degrees) no (forgotten) do not reject it (from)	
true (what) .485 (scholar) reject it (you) false (learned) 5.991 (have)	
2 (you) 1.697 (what) 33 (subtract) 29 (school) 99 (knowledge)	

Write the answer to the riddle here, putting one word on each line: _____ _____ _____ _____
_____ _____ _____ _____ _____

Section 50: Computations for a Two-Way Chi Square

Suppose that three randomly selected groups were drawn from a population of people suffering from a chronic disease. One group was administered a large dose of an experimental drug, another was administered a small dose of the drug, and the third group, which served as a control, was administered a placebo that looked like the drug but was inert. Subjects were also classified as to whether their condition was improved. The following table shows the **observed frequencies** that resulted:[1]

Table 50.1
Observed Frequencies for Three Treatment Groups

	Large Dose	Small Dose	Placebo	**Row Totals**
Improved	20	15	9	**44**
Not Improved	10	14	20	**44**
Column Totals	**30**	**29**	**29**	**Grand Total = 88**

Inspection of the frequencies suggests that the large dose was more effective than the small dose and that both dosage levels were more effective than the placebo. Because only random samples were observed, however, we need to test the null hypothesis.

Before using the formula for chi square, we need to calculate the **expected frequencies**. Here is the rule for calculating expected frequencies for a two-way chi square:

For each cell, multiply the associated row total by the associated column total and divide by the grand total.

The next table shows the application of this rule.

---

[1]Normally, an investigator would assign the same number of subjects to each treatment group in an experiment of this type. However, some subjects may drop out, resulting in unequal numbers of subjects for analysis.

Some students have trouble mastering this rule at first, so let us take it slowly. Consider the cell in Table 50.1 for the subjects who had the *large dose* and were *improved*. This cell is at the upper left and has an *observed frequency* of 20. The row total associated with this cell is 44 (it is shown at the right) and the column total associated with this cell is 30 (it is shown at the bottom). To apply the rule, multiply 44 by 30 and divide by the grand total of 88. This computation is shown in the upper left of Table 50.2. The computations for all the other cells are also shown.

Table 50.2
Computation of Expected Frequencies

	Large Dose	Small Dose	Placebo
Improved	$\dfrac{(44)(30)}{88} = \mathbf{15.000}$	$\dfrac{(44)(29)}{88} = \mathbf{14.500}$	$\dfrac{(44)(29)}{88} = \mathbf{14.500}$
Not Improved	$\dfrac{(44)(30)}{88} = \mathbf{15.000}$	$\dfrac{(44)(29)}{88} = \mathbf{14.500}$	$\dfrac{(44)(29)}{88} = \mathbf{14.500}$

Now we can apply the formula for chi square, which is:

$$\chi^2 = \Sigma \frac{(O-E)^2}{E}$$

When applying the formula, you must apply it separately for each cell—in this case, we have 6 cells. For the cell for those who improved with the large dose (the upper left), the observed frequency (O) from Table 50.1 is 20 and the expected frequency (E) from Table 50.2 is 15.000. These values are entered into the formula here:

For Improved/Large Dose: $\dfrac{(20-15.000)^2}{15.000} = \dfrac{(5.000)^2}{15.000} = \dfrac{25.000}{15.000} = \mathbf{1.667}$

Then do the same thing for each of the other cells. Make sure that you are always looking at corresponding cells to get the observed and expected frequencies. These are the computations:

For Improved/Small Dose: $\dfrac{(15 - 14.500)^2}{14.500} = \dfrac{(.500)^2}{14.500} = \dfrac{.250}{14.500} = \textbf{.017}$

For Improved/Placebo: $\dfrac{(9 - 14.500)^2}{14.500} = \dfrac{(-5.500)^2}{14.500} = \dfrac{30.250}{14.500} = \textbf{2.086}$

For Not Improved/Large Dose: $\dfrac{(10 - 15.000)}{15.000} = \dfrac{(-5.000)^2}{15.000} = \dfrac{25.000}{15.000} = \textbf{1.667}$

For Not Improved/Small Dose: $\dfrac{(14 - 14.500)^2}{14.500} = \dfrac{(-.500)^2}{14.500} = \dfrac{.250}{14.500} = \textbf{.017}$

For Not Improved/Placebo: $\dfrac{(20 - 14.500)^2}{14.500} = \dfrac{(5.500)^2}{14.500} = \dfrac{30.250}{14.500} = \textbf{2.086}$

Summing the six values computed above yields the value of chi square:

$$\chi^2 = 1.667 + .017 + 2.086 + 1.667 + .017 + 2.086 = \textbf{7.540}$$

Thus, **7.540** is our *observed value of chi square*.

Before we can determine significance, we need to compute the degrees of freedom. For a two-way chi square, the formula is:

df = (number of rows – 1)(number of columns – 1)

Since we have two rows (improved and not improved) and three columns (large dose, small dose, and placebo), the degrees of freedom for our example are:

df = (2 – 1)(3 – 1) = (1)(2) = **2**

To determine significance, compare the *observed value of chi square* that we computed with the *critical value of chi square* in Table 11. In this Table, we find that for 2 degrees of freedom, the critical value at the .05 level is **5.991**.

Here is our decision rule:

If the *observed value of chi square is greater than the critical value,*[2] reject the null hypothesis; otherwise, do not reject it.

Since the observed value (7.540) is greater than the critical value (5.991), we should reject the null hypothesis at the .05 level. We should conclude that the differences are statistically significant because we have ruled out random sampling error as the explanation for them.

One note of caution: The probabilities obtained with chi square will not be accurate if the expected frequency in any cell is small, that is, less than about 10. In most cases, this can be prevented by using a reasonably large sample size. Another solution is to "collapse" adjoining cells when there is a logical basis for doing so. For example, suppose you obtained the following expected frequencies for an experiment:

	Experimental Group	Control Group
Major Improvement	4	1
Minor Improvement	12	10
No Improvement	12	15

By collapsing (i.e., combining) the frequencies for Major Improvement with Minor Improvement, the expected frequencies for the analysis can be made larger.[3]

---

[2]In the unlikely event that the observed value is exactly equal to the critical value, reject the null hypothesis.

[3]Several modifications to the chi square formula have been suggested for use when one or more expected frequency is small. These are not universally accepted among statisticians and only apply, in any case, to problems with one degree of freedom. Thus, the solutions suggested above are more satisfactory. One of the most popular modifications is the Yates' correction, which you may encounter when you read journal articles—especially older ones.

Terms to Review Before Attempting Worksheet 50

**observed frequencies, expected frequencies,
observed value of chi square, critical value of chi square**

"I couldn't do my statistics homework because my
computer has a virus and so do all my pens and pencils."

Worksheet 50: Computations for a Two-Way Chi Square

> *Riddle*: **What proves that people are wrong to complain about the postal service?**

DIRECTIONS: To find the answer to the riddle, write the answer to each question in the space immediately below it. The word in parentheses in the solution section next to the answer to the first question is the first word in the answer to the riddle, the word beside the answer to the second question is the second word, and so on.

A random sample of subjects from a community was asked whether they favor or oppose welfare. They were also asked whether they had ever been on welfare. The following frequencies were obtained.

	Favor Welfare	Oppose Welfare
Had Been on Welfare	15	10
Never on Welfare	50	60

1. Is someone who "had been on welfare" *or* someone "never on welfare" more likely to favor welfare?

2. What is the grand total of the frequencies?

3. What is the expected frequency for the Never on Welfare/Favor Welfare cell?

4. What is the expected frequency for the Never on Welfare/Oppose Welfare cell?

Worksheet 50 (Continued)

5. To three decimal places, what is the observed value of chi square? (Allow for minor variations in the third place due to rounding.)

6. What is the value of the degrees of freedom?

7. What is the critical value of chi square at the .05 level?

8. Is chi square significant at the .05 level?

9. "The null hypothesis should be rejected at the .05 level." Is this statement true or false?

Solution section:

4 (slowly)	5.991 (delivery)	false (time)	no (on)	3.841 (always)
1 (are)	1.726 (bills)	never on welfare (phone)	had been on welfare (postage)	
135 (bills)	52.963 (and)	57.037 (electric)	2 (complain)	3 (left)

Write the answer to the riddle here, putting one word on each line: _____ _____ _____ _____
_____ _____ _____ _____

315

Notes:

Section 51: Cramér's Phi

A two-way chi square (see Sections 48 and 50) tests whether there is a statistically significant relationship between two variables. In Section 50, we tested the significance of the relationship between the two variables shown in Table 51.1.

Table 51.1
Observed Frequencies for Three Treatment Groups

	Large Dose	Small Dose	Placebo	**Row Totals**
Improved	20	15	9	**44**
Not Improved	10	14	20	**44**
Column Totals	**30**	**29**	**29**	**Grand Total = 88**

The observed frequencies in Table 51.1 suggest that there is a relationship between (1) improvement/no improvement and (2) amount of drug taken. The larger the amount taken, the more improvement. This relationship was found to be statistically significant as indicated by the fact that:

$$\chi^2 = 7.540, p < .05$$

Thus, we can reject the null hypothesis, which says that the relationship was created by sampling errors.

Note, however, that a significance test does not tell us how strong a relationship is. In this case, the chi square test of significance only tells us that the relationship is reliable (i.e., significant), but even a weak relationship is sometimes reliable. Inspection of Table 51.1 clearly indicates that the relationship we are considering is far from perfect. For example, while 20 of the 30 who took the large dose (about 67%) improved, 9 of the 29 who received the placebo (about 31%) also improved. Thus, improvement is not entirely contingent on receiving the drug; the relationship between improvement and dosage level is, therefore, far less than perfect.

Cramér developed a statistic based on the Pearson r that describes the *strength* of the relationship between two variables examined in a two-way chi square.[1] Here's the formula for ***Cramér's phi***, whose symbol is ϕ.

$$\phi = \sqrt{\frac{\chi^2}{N(k-1)}}$$

Where:

χ^2 is the value of chi square.

N is the total number of cases (the grand total).

k is the number of categories for the variable with the smaller number of categories. (In our example, there are 2 categories of improvement [improved and not improved] and 3 categories of drug usage [large dose, small dose, and placebo]; thus, $k = 2$ for our example.)[2]

Solving our example, we obtain:

$$\phi = \sqrt{\frac{7.540}{(88)(2-1)}} = \sqrt{.0857} = .2927 = .29$$

To interpret Cramér's phi, keep in mind that 0.00 indicates no relationship and 1.00 indicates a perfect relationship. Thus, for our example, phi indicates that the relationship is weak.

When reporting our results, we can now indicate that there is a statistically significant relationship as indicated by chi square but the relationship is weak as indicated by Cramér's phi. By reporting the value of phi along with the results of the chi square test, we give consumers of our research a more comprehensive report of results than if we had just reported the results of chi square.

---

[1]Remember that to compute a Pearson r, you need two sets of scores (see Section 21). Here, we are dealing with frequencies—not scores. It is possible to arbitrarily assign scores to the categories and compute a Pearson r as a measure of relationship. Cramér's method, however, is much easier if a chi square test has already been computed.

[2]If both variables have the same number of categories, then use the number in either one as k. For example, if there are two categories on variable A and two on variable B, use 2 as the value of k.

Term to Review Before Attempting Worksheet 51

Cramér's phi

**"Class, who can tell me what I have preserved in this jar?
No, it's not a pig or a baby cow...it's the last student
who got caught cheating on one of my tests!"**

Worksheet 51: Cramér's Phi

Riddle: Why are people who live in the lap of luxury anxious?

DIRECTIONS: To find the answer to the riddle, write the answer to each question in the space immediately below it. The word in parentheses in the solution section next to the answer to the first question is the first word in the answer to the riddle, the word beside the answer to the second question is the second word, and so on.

1. If there is no relationship between two variables, what will be the value of Cramér's phi?

2. If a relationship is statistically significant, is it necessarily a strong relationship?

Questions 3 through 5 refer to the information in the following table on gender preferences for three candidates. In the cells are frequencies (i.e., numbers of cases).

	Candidate A	Candidate B
Males	20	30
Females	30	20
$\chi^2 = 4.000$, $df = 1$, $p < .05$		

3. What is the value of N for use in the formula for Cramér's phi?

4. What is the value of k for use in the formula for Cramér's phi? (See footnote 2 on page 318.)

320

Worksheet 51 (Continued)

5. To two decimal places, what is the value of Cramér's phi?

Questions 6 through 8 refer to the information in the following box.

> Two hundred patients were randomly assigned to one of three treatments for a life-threatening disease. After 5 years, they were classified as either alive or dead. The relationship between type of treatment and survival was statistically significant ($\chi^2 = 25.991$, $df = 2$, $p < .001$).

6. What is the value of N for use in the formula for Cramér's phi?

7. To two decimal places, what is the value of Cramér's phi?

8. As indicated by Cramér's phi, is the relationship "strong" *or* "weak"?

Solution section:

> 1.00 (wealth) .20 (might)   200 (suddenly) .36 (stand) .40 (but)
>
> 0.00 (they're) 100 (that) strong (lift) yes (golden) no (afraid)
>
> 2 (luxury) weak (up) .04 (dependent) 50 (something)

Write the answer to the riddle here, putting one word on each line: _____ _____ _____ _____

_____ _____ _____ _____

Notes:

Section 52: Median Test

Although the mean is the most frequently used average, it is sometimes inappropriate.[1] The most popular alternative is the median—the value that has half the scores above it and half the scores below it. In Section 11, you learned how to compute the median.[2] A common problem is when we have sampled at random from two independent groups, computed the median for each group, and want to compare the two medians for statistical significance. The *median test* may be used for this purpose. Here is an example:

Example 1:

A random sample of subjects was drawn from the east side of town and a random sample was drawn from the west side. They were asked to indicate their annual incomes to the nearest thousand dollars. These results were obtained (in thousands of dollars, for example 5 = $5,000.00):

East Side	West Side
5	6
7	9
8	12
10	14
17	20
30	40
33	45
41	55
44	61
median = 17	median = 20

[1]The mean is inappropriate when a distribution of scores is highly skewed, in which case the median is preferred. Also, if the data are ordinal, the mean is inappropriate and the median should be used. See Section 11 to review these concepts.

[2]The procedure in Section 11 is only approximate if there are ties (multiple scores with the same value) at the midpoint. Appendix H illustrates how to calculate the median when there are ties.

The medians suggest that those who live on the west side, on the average, have higher incomes than those on the east side. However, only random samples have been drawn, therefore, forcing us to consider the null hypothesis, which states that there is no *true* difference. To test this null hypothesis, we calculate chi square. To do so, follow these steps:

Step 1: Arrange *all* the scores in order from low to high.

5, 6, 7, 8, 9, 10, 12, 14, 17, 20, 30, 33, 40, 41, 44, 45, 55, 61

Step 2: Determine the median of the scores listed in Step 1.

Because there are 18 scores, count up 9 (half of 18) scores to reach the middle; we reach a score of 17. Count down 9 scores to reach the middle; we reach a score of 20. Thus, the median is halfway between 17 and 20 ($17 + 20 = 37/2 = 18.5$). The *median of all the scores* is **18.5**.

Step 3: Determine how many subjects in *each* group are above and below the *median of all the scores* (i.e., 18.5) and arrange them in a table as shown below.[3] For example, among those on the east side, there are 4 subjects with incomes above 18.5.

Number above and below 18.5 in each group.		
	East Side	West Side
Number *above* median	4	5
Number *below* median	5	4

Step 4: Conduct the usual two-way chi square test using the method described in Section 50. Applying the method, this result is obtained:

$\chi^2 = .222$, $df = 1$, not significant at the .05 level.

---

[3]If a subject's score is exactly equal to the median, include the subject in the *below* median group. Do not use this test if there are 2 or fewer subjects in any group in your table.

Because the value of chi square in our example is *not* significant, we do *not* reject the null hypothesis and we do *not* declare statistical significance.

In the next section, you will be introduced to an alternative to the median test.

Term to Review Before Attempting Worksheet 52

median test

**"According to the latest statistics, the average
human body is 20% water and 80% stress."**

Worksheet 52: Median Test

> **Riddle:** What does an addicted shopper do at the end of the day?

DIRECTIONS: To find the answer to the riddle, write the answer to each question in the space immediately below it. The word in parentheses in the solution section next to the answer to the first question is the first word in the answer to the riddle, the word beside the answer to the second question is the second word, and so on.

The questions refer to these scores. You will need to use the technique for conducting a two-way chi square described in Section 50.

Group A	Group B
0	3
1	4
5	6
11	17
12	18
16	29
23	35
33	39
44	55
median = 12	median = 18

1. On the average, which group has higher scores?

2. What is the median of *all* the scores?

Worksheet 52 (Continued)

3. How many subjects in Group A are below the median of *all* the scores?

4. How many subjects in Group B are below the median of *all* the scores?

5. What is the value of chi square?

6. Is chi square statistically significant at the .05 level?

7. "The null hypothesis may be rejected at the .05 level." Is this statement true or false?

Solution section:

| Group A (buys) Group B (leaves) 16.50 (the) 6 (mall) 4 (runs) |
| yes (always) .50 (cash) 3 (with) no (but) 16.00 (credit) |
| 2.00 (everything) 18 (fashions) false (money) 15.00 (bankrupt) |

Write the answer to the riddle here, putting one word on each line: _____ _____ _____ _____ _____ _____ _____

Notes:

Section 53: Mann-Whitney *U* Test

The ***Mann-Whitney U* test** is an alternative to the median test presented in the previous section.[1] It tests whether the distribution of scores for one random sample is significantly different from the distribution of another independent random sample.[2]

The test will be illustrated using the following scores, which were also analyzed in the previous section:

Col. 1	Col. 2
East Side	West Side
(Group 1)	(Group 2)
5	6
7	9
8	12
10	14
17	20
30	40
33	45
41	55
44	61
median = 17	median = 20

Step 1: Rank all scores as though the two groups were a single group.

When ranking, give a rank of 1 to the lowest score, a rank of 2 to the next lowest score, etc.[3]

---

[1]The Mann-Whitney *U* test is more powerful than the median test; that is, it is more likely to lead to the rejection of the null hypothesis than the median test.

[2]If the two distributions have similar shapes, when significance is obtained with the Mann-Whitney *U* test, it is safe to assume that the two medians are significantly different.

[3]If there are ties, split the ranks among them. For example, if the two lowest scores are 4 and 4 (i.e., two subjects have a 4), they are tied and it is not possible to determine which should have a rank of 1 and which should have a rank of 2. We compromise by adding the ranks in question (1 + 2 = 3) and dividing by the number of scores that are tied (in this case, two scores). Thus, 3/2 = 1.5, which is the rank assigned to each one.

Although all scores were ranked as though they were a single group, list the ranks separately by groups in Columns 3 and 4 as shown below.

Col. 1	Col. 2	Col. 3	Col. 4
East Side	West Side	Ranks for	Ranks for
(Group 1)	(Group 2)	Group 1	Group 2
5	6	1	2
7	9	3	5
8	12	4	7
10	14	6	8
17	20	9	10
30	40	11	13
33	45	12	16
41	55	14	17
44	61	15	18
median = 17	median = 20	$\Sigma R_1 = 75$	$\Sigma R_2 = 96$
$n_1 = 9$	$n_2 = 9$		

Step 2: Sum the ranks in Columns 3 and 4 and enter them at the bottom of the columns as shown above. For example, the sum of the ranks for Group 1 is $\Sigma R_1 = 75$, where R stands for the ranks.

Step 3: Solve for the value of U_1 for Group 1 using this formula:

$$U_1 = (n_1)(n_2) + \frac{n_1(n_1 + 1)}{2} - \Sigma R_1$$

$$= (9)(9) + \frac{9(9 + 1)}{2} - 75$$

$$= 81 + \frac{90}{2} - 75 = 81 + 45 - 75 = \mathbf{51}$$

Step 4: Solve for the value of U_2 for Group 2 using this formula:

$$U_2 = (n_1)(n_2) + \frac{n_2(n_2 + 1)}{2} - \Sigma R_2$$

$$= (9)(9) + \frac{9(9 + 1)}{2} - 96$$

$$= 81 + \frac{90}{2} - 96 = 81 + 45 - 96 = \mathbf{30}$$

Step 5: Inspect the results of Steps 3 and 4 to determine which value of U is *smaller*. (That is, determine whether U_1 or U_2 is smaller.)

In our example, U_2, which equals **30**, is smaller than U_1.

Step 6: Evaluate the value of U identified in Step 5 using Table 12 near the end of this book for the .05 level to determine significance.[4] To find the critical value, identify where the row for N_1 meets the column for N_2. For our example, the *critical value of U* is **17**.

Here is our decision rule:

> If the *observed value of U is less than the critical value,*[5]
> reject the null hypothesis; otherwise, do not reject it.

Because our observed value (**30**) is *not* less than the critical value (**17**), we do not reject the null hypothesis, and we do *not* declare the difference to be statistically significant at the .05 level.

A note of caution: The Mann–Whitney U test, as illustrated in this section, should be used only when $n = 20$ or less for the larger of the two groups *and n*

---

[4]Use Table 13 for the .01 level. Tables 12 and 13 are for two-tailed tests.
[5]Be careful: In all of our previous decision rules, the observed value had to be *greater* than the critical value in order to declare significance. Also, in the unlikely event that the observed value is exactly equal to the critical value, reject the null hypothesis.

= 9 or more for the larger group. Application of the test for situations beyond these parameters is beyond the scope of this book.

Term to Review Before Attempting Worksheet 53

Mann-Whitney *U* test

"How to Lose Weight While Doing Statistics Homework: Double-click mouse six million times between each data entry."

Worksheet 53: Mann-Whitney *U* Test

| Riddle: According to R. E. Shay, what proves that a rabbit's foot is not lucky? |

DIRECTIONS: To find the answer to the riddle, write the answer to each question in the space immediately below it. The word in parentheses in the solution section next to the answer to the first question is the first word in the answer to the riddle, the word beside the answer to the second question is the second word, and so on.

Col. 1	Col. 2	Col. 3	Col. 4
East Side	West Side	Ranks for Group 1	Ranks for Group 2
(Group 1)	(Group 2)		
1	0		
3	4		
5	6		
11	17		
12	18		
16	29		
23	35		
33	39		
44	55		
median = 12	median = 18	$\Sigma R_1 =$	$\Sigma R_2 =$
$n_1 = 9$	$n_2 = 9$		

1. What is the rank for a person with a score of 18?

2. What is the value of *U* for Group 1?

Worksheet 53 (Continued)

3. What is the value of U for Group 2?

4. Should the value of U for "Group 1" *or* the value for "Group 2" be used as the observed value for the test of significance?

5. What is the critical value of U at the .05 level?

6. Should the null hypothesis be rejected at the .05 level?

7. "The difference is statistically significant at the .05 level." Is this statement true or false?

Solution section:

Group 1 (fur) yes (fortune) true (being) 11 (it) 49 (did) 32 (not)
8 (animal) false (rabbit) no (the) 17 (for) 7 (money)
14 (windows) Group 2 (work) 15 (charm) 10 (superstitious)

Write the answer to the riddle here, putting one word on each line: _____ _____ _____ _____ _____ _____ _____

Section 54: Wilcoxon's Matched-Pairs Test

The median test in Section 52 and the Mann-Whitney U test in Section 53 are used with independent data. *Wilcoxon's matched-pairs test* (also known as the *matched-pairs signed-ranks test*) is for use when one subject in each group is matched with one subject in the other group.[1] An example of matched pairs follows:

Fourteen students were selected at random for an experiment on learning advanced basketball skills. Students were paired by matching them on the results of a test of elementary basketball skills; that is, the two subjects with the lowest elementary skills were designated as one pair, the two with the next lowest elementary skills were designated as another pair, etc. For each pair, a coin was flipped to determine which one would become a member of the experimental group; the other member of the pair became a member of the control group. These are the scores they earned on a test of advanced skills at the end of the experiment:

Pair	Col. 1 Experimental	Col. 2 Control
A	3	6
B	7	1
C	9	8
D	15	13
E	26	19
F	30	22
G	35	31
H	20	9
I	16	6
	Median = 16	Median = 9

---

[1]Thus, Wilcoxon's matched-pairs test is for *dependent* or *correlated* data while the tests in Sections 52 and 53 are for *independent* or *uncorrelated* data. All three tests are used with ordinal data or data that violate the assumptions of the *t* test or ANOVA.

Inspection of the medians suggests that the performance of the experimental group was superior to that of the control group. However, because only a random sample was studied, we are forced to consider the null hypothesis, which states that there is no *true* difference between the distributions. To test the null hypothesis, follow these steps:

Step 1: Calculate the *absolute* difference between each pair of scores. (The *absolute difference* is the difference without regard to sign. In practical terms, this means to record all differences as though they were positive; do not record any negative signs.) Thus, subtract each score in Column 2 from the other member of the pair in Column 1, ignore the sign of the answer, and enter the absolute differences in Column 3, as shown in the following table:

Pair	Col. 1 Experimental	Col. 2 Control	Col. 3 Absolute Difference
A	3	6	3
B	7	1	6
C	9	8	1
D	15	13	2
E	26	19	7
F	30	22	8
G	35	31	4
H	20	9	11
I	16	6	10

Step 2: Rank the differences in Column 3 and record the ranks in Column 4. When ranking, give the smallest difference a rank of 1, the next smallest difference a rank of 2, etc.[2] The ranks are shown in the following table.

---

[2]If there are ties, split the ranks among them. For example, if the two lowest scores are 4 and 4 (i.e., two subjects have a 4) they are tied and it is not possible to determine which should have a rank of 1 and which should have a rank of 2. We compromise by adding the ranks in question (1 + 2 = 3) and dividing by the number of scores that are tied (in this case, two scores). Thus, 3/2 = 1.5, which is the rank assigned to each one.

Pair	Col. 1 Experimental	Col. 2 Control	Col. 3 Absolute Difference	Col. 4 Rank
A	3	6	3	3
B	7	1	6	5
C	9	8	1	1
D	15	13	2	2
E	26	19	7	6
F	30	22	8	7
G	35	31	4	4
H	20	9	11	9
I	16	6	10	8

Step 3: Record in Column 5 whether the difference between Columns 1 and 2 (subtracting Col. 2 from Col. 1 in each case) *would have been* positive or negative. This is done in Column 5 in the table below.

Pair	Col. 1 Exp.	Col. 2 Control	Col. 3 Absolute Difference	Col. 4 Rank	Col. 5 Positive/ Negative
A	3	6	3	3	—
B	7	1	6	5	+
C	9	8	1	1	+
D	15	13	2	2	+
E	26	19	7	6	+
F	30	22	8	7	+
G	35	31	4	4	+
H	20	9	11	9	+
I	16	6	10	8	+

Step 4: Record the *ranks* associated with the positive signs in Column 6 and the *ranks* associated with the negative signs in Column 7 as shown in the table below. Then sum these two columns.

	Col. 1	Col. 2	Col. 3	Col. 4	Col. 5	Col. 6	Col. 7
Pair	Exp.	Control	Absolute Difference	Rank	Pos./ Neg.	Rank+ $R+$	Rank– $R-$
A	3	6	3	3	–		3
B	7	1	6	5	+	5	
C	9	8	1	1	+	1	
D	15	13	2	2	+	2	
E	26	19	7	6	+	6	
F	30	22	8	7	+	7	
G	35	31	4	4	+	4	
H	20	9	11	11	+	9	
I	16	6	10	10	+	8	
						$\Sigma R+ = 42$	$\Sigma R- = 3$

Step 5: Determine whether $\Sigma R+$ or $\Sigma R-$ is smaller. The smaller one is the *observed value* of Wilcoxon's T statistic. Therefore, for our example:

Wilcoxon's $T = 3$

Step 6: Determine the *critical value* of Wilcoxon's T using Table 14 near the end of this book, using this formula for degrees of freedom:[3]

$$df = N_{\text{pairs}} - 1 = 9 - 1 = 8$$

Table 14 indicates that for 8 degrees of freedom at the .05 level, the *critical value* is **4**.

---

[3]This table is for a two-tailed test.

Here is our decision rule:

> If the *observed value of Wilcoxon's T is less than the critical value*,[4] reject the null hypothesis.

Since our *observed value* (**3**) is less than the *critical value* (**4**), we may reject the null hypothesis at the .05 level and declare the difference to be statistically significant.

Terms to Review Before Attempting Worksheet 54

Wilcoxon's matched-pairs test (matched-pairs signed-ranks test)

"I forgot to make a back-up copy of my brain, so everything I learned in statistics last semester was lost."

---

[4]In the unlikely event that the observed value is exactly equal to the critical value, reject the null hypothesis.

Worksheet 54: Wilcoxon's Matched-Pairs Test

> **Riddle:** According to the *Devil's Dictionary*, what is the definition of love?

DIRECTIONS: To find the answer to the riddle, write the answer to each question in the space immediately below it. The word in parentheses in the solution section next to the answer to the first question is the first word in the answer to the riddle, the word beside the answer to the second question is the second word, and so on.

Pair	Col. 1 Exp.	Col. 2 Control	Col. 3 Absolute Difference	Col. 4 Rank	Col. 5 Pos./ Neg.	Col. 6 Rank+ $R+$	Col. 7 Rank− $R−$
A	10	3					
B	8	6					
C	7	10					
D	1	12					
E	6	2					
F	15	5					
G	10	9					
H	10	5					
						$\Sigma R+ =$	$\Sigma R− =$

1. What is the rank for Pair A?

2. What is the sum of Column 6?

Worksheet 54 (Continued)

3. What is the sum of Column 7?

4. What are the degrees of freedom?

5. What is the critical value of Wilcoxon's T at the .05 level?

6. May the null hypothesis be rejected at the .05 level?

7. "The difference is statistically significant at the .05 level." Is this statement true or false?

Solution section:

8 (right) 4 (bliss) 25 (is) 6 (love) 11 (temporary) false (marriage)
true (feelings) no (by) 2 (curable) yes (right) 7 (insanity)
15 (flowers) 20 (proof) 16 (wedding) 0 (heart) 1 (nothing)

Write the answer to the riddle here, putting one word on each line: _____ _____ _____ _____,
_____ _____ _____

Notes:

Section 55: Descriptive Statistics: Their Value in Research

In Part A of this book (Sections 1 through 26), you have learned about a wide variety of *descriptive statistics*. They are of great value for three reasons. First, they help researchers obtain overviews of selected characteristics of the groups they study. Second, they make it relatively easy to compare two or more groups on the characteristics (i.e., variables) being studied. Finally, they allow researchers to communicate their findings quickly and efficiently with others. These three reasons are illustrated in the following examples.

Example 1:

A medical researcher assigned 180 patients with a potentially terminal illness to two groups at random. The experimental group ($n = 90$) received a new drug designed to treat the illness, while the other group ($n = 90$) received a placebo that looked and tasted like the new drug but was actually inert. At the end of the study, the researcher had collected these data:

Participant	Group	Outcome
Jill	Experimental	Alive
Fernando	Control	Dead
Suzanne	Control	Dead
Rachel	Experimental	Dead
Roxanne	Control	Dead
Wayne	Experimental	Alive

Plus additional results of the type shown above for the remaining 174 participants.

The researcher in Example 1 collected **nominal** data (see Section 2 of this book). It is *nominal* because, unlike scores, the data are *names* such as "experimental" and "control" as well as "alive" and "dead." The researcher can use descriptive statistics to count the **frequencies** (see Section 3) of those who are alive and dead in each group. (The symbol for frequencies is f. However, frequencies are usually reported in research reports as *numbers of participants*, whose symbol is n. In other words, f and n are equivalent for all practical purposes.) The researcher can also compute the **percentages** or **proportions** (see Section 3) that correspond to the frequencies. These can be reported in sentences or a statistical table such as this one:

| | Outcome | |
	Dead	Alive
Experimental Group	42%	58%
	($n = 38$)	($n = 52$)
Control Group	74%	26%
	($n = 67$)	($n = 23$)

The descriptive statistics in the table of results shown immediately above (1) provide an overview of the results, (2) make it easy to compare the outcome for the experimental group with the outcome for the control group, and (3) provide an efficient way to communicate the results of the experiment to others. Thus, Example 1 illustrates the three reasons that make descriptive statistics exceptionally useful in research.

Example 2:

In May of each year, administrators test the ability of all first-grade students in a school district to add and subtract one-digit numbers using a 20-item completion test with questions such as "3 + 4 = ___." For each of the 325 first graders in the three schools in the district this year, they obtained a number-right score, the name of the school each student attends, and the name of the class in which the students are enrolled identified by teachers' names (e.g., Marlyn = 14 right, School X, Teacher A). This is the beginning of the list of the 325 results in no particular order (i.e., not yet analyzed with descriptive statistics):

Student	Number Right	School	Teacher
Mary	15	Washington	Jones
Jose	19	Franklin	Black
Hilary	17	Bishop	Doe
Francis	21	Washington	Smith
Tom	8	Bishop	Doe
Smitty	3	Washington	Jones

Plus additional results of the type shown above for the remaining 319 participants.

As you can see, the results in Example 2 need to be organized so that the school administrators can get an overview, compare groups (such as comparing the various schools with each other), and communicate the results to interested parties such as parents and the school board. While there are various ways to do this using descriptive statistics, the following steps are standard for these types of data:

Step 1: Determine the *scale of measurement* (see Section 2). Note that most researchers treat multiple-choice test scores as being at the *interval scale of measurement*. This is done based on the assumption that each correct answer on such a test represents about the same amount of knowledge as each other correct answer. Thus, researchers assume, for example, that the interval between 10 right and 11 right is about the same size (in terms of amount of knowledge) as the interval between 11 right and 12 right (i.e., that data points generated with multiple-choice tests have *equal intervals*). Note that if the scores only ranked students (and did not measure with equal units), the scale would be *ordinal* (see Section 2). Also, note that the assumption that the scores are at equal intervals will influence the selection of additional statistics in subsequent steps below.

Step 2: Examine the shape of the distribution using a statistical table such as a *frequency distribution* (see Sections 4 and 5) and/or by constructing a statistical figure such as a ***histogram*** (see Section 7) or a ***frequency polygon*** (see Section 8).

Step 3: If the distribution is at least roughly symmetrical and not highly ***skewed*** (see Section 9), then select the ***mean*** as the average (see Section 10). If the scale of measurement is ordinal such as ***ranks*** (see Section 2 and Step 1 above) or if the distribution is highly skewed, select the ***median*** as the average (see Section 11). Compute the average (i.e., *mean* or *median*) for all 325 students. Also, compute the average for each group the administrators want to compare, such as the students at each of the three schools (Washington, Franklin, and Bishop), so that the averages can be compared across schools.

Step 4: If the administrators also want to use the test scores to determine how each student performed in relation to the other students in the school or the district, compute ***percentile ranks*** (see Section 6) and/or ***standard scores*** (see Sections 16 and 17) for each student.

Step 5: Examine the variability in the scores. If the median was selected as the average, compute the ***range*** and ***interquartile range*** (see Section 12) as the measures of variability. If the mean was selected as the average, compute the ***standard deviation*** (see Sections 13 through 15). Note that important information can be obtained by comparing a measure of variability across schools and classes. For instance, if one school has a much larger standard deviation (i.e., students tend to be further from the mean in one school) than the other schools, the instructional techniques used in that school might need to be modified for that school. For example, when there is a wide variation in a group of students (as indicated, for instance, by a very large standard deviation), individualized instruction is probably preferable to whole-class instruction such as group lectures.

By following the five steps listed above, the school administrators will be able to (1) obtain an overview of the students' abilities to add one-digit numbers—either at the district, school, or classroom level, (2) compare various groups of students within the school district, and (3) concisely communicate important information such as individual students' percentile ranks as well as information about the groups' averages and variability.

Consider Example 3, in which there is one set of scores for two groups of research participants.

Example 3:

A research psychologist developed a new scale to measure depression, which consists of 30 true–false statements such as "I often feel sad when I wake up in the morning." The scale yields scores that can range from zero to 30. Both theory and research on depression indicate that anxiety is often associated with depression, which suggests that there should be a modest relationship between the two variables. Therefore, the psychologist administered the scale to 75 research participants. To the same group of participants, the psychologist administered a previously published, standardized measure of anxiety that yields *T scores* (see Section 17). The *T* scores can range from 20 to 80. The following data were obtained:

Participant	Depression Score	Anxiety Score
Jose	29	55
Sally	5	25
Ling	15	45
Sammy	10	75

Plus additional results of the type shown above for the remaining 71 participants.

The research psychologist in Example 3 determined that the distributions of the two sets of scores are not highly *skewed* (see Section 9) by examining the *frequency distribution for grouped data* (see Section 5) and a *frequency polygon* (see Section 8), and assumed that both variables are measured at the *interval scale* (see Section 2). The psychologist calculated the **mean** for each variable (see Sections 10 and 11) and the **standard deviation** (see Sections 13 through 15). Then, the psychologist constructed a **scattergram** (see Section 20) and determined that the relationship is reasonably close to being **linear** (see Section 20). However, the psychologist observed considerable scatter in the scattergram, which indicates a far less than perfect relationship between the two sets of scores. To obtain a single numerical value to represent this relationship, the psychologist computed the **Pearson r** (see Sections 21 and 22) and obtained an r of 0.44. To interpret this value, the psychologist computed the **coefficient of determination** (see Section 23), which is the square of r (0.44 × 0.44 = 0.19). Multiplying the coefficient of determination by 100% (0.19 × 100% = 19%), the psychologist determined that the amount of variance on anxiety accounted for by her new measure of depression is 19%, which indicates that her new measure operates as predicted by the theory and previous research on these variables (i.e., anxiety and depression are modestly related). The psychologist wrote a journal article that reported on the development of the new measure of depression. The article included a discussion of the descriptive statistics mentioned above. These provided readers with a concise overview, obviating the need to report the two scores for each of the 75 individual participants.

Terms to Review Before Attempting Worksheet 55

nominal (see Section 2), **frequencies** (see Section 3), **percentages** or **proportions** (see Section 3), **scale of measurement** (see Section 2), **ordinal** (see Section 2), **frequency distribution** (see Sections 4 and 5), **histogram** (see Section 7), **frequency polygon** (see Section 8), **skewed** (see Section 9), **mean** (see Section 10), **ranks** (see Section 2), **median** (see Section 11), **percentile ranks** (see Section 6), **standard scores** (see Sections 16 and 17), **range** (see Section 12), **interquartile range** (see Section 12), **standard deviation** (see Sections 13–15), **T scores** (see Section 17), **skewed** (see Section 9),

frequency distribution for grouped data (see Section 5),
frequency polygon (see Section 8), **interval scale** (see Section 2),
mean (see Sections 10 and 11), **standard deviation** (see Sections 13–15),
scattergram (see Section 20), **linear** (see Section 20),
Pearson *r* (see Sections 21 and 22), **coefficient of
determination** (see Section 23)

Englebert Working on His Statistics Homework

Worksheet 55: Descriptive Statistics: Their Value in Research

---

Riddle: A "clear conscience" is usually a sign of what?

DIRECTIONS: To find the answer to the riddle, write the answer to each question in the space immediately below it. The word in parentheses in the solution section next to the answer to the first question is the first word in the answer to the riddle, the word beside the answer to the second question is the second word, and so on.

1. According to the information in this section, descriptive statistics are of great value for how many reasons?

2. The researcher in Example 1 collected data that were at what scale of measurement ("nominal," "ordinal," *or* "interval")?

3. What is the symbol for number of participants?

4. In Example 1, which group ("experimental" *or* "control") had a higher percentage of participants who had died by the end of the study?

5. Most researchers treat multiple-choice test scores as being at what level (scale) of measurement?

6. If the scale of measurement is interval and if the distribution is at least roughly symmetrical and not highly skewed, which average should be selected?

350

Worksheet 55 (Continued)

7. To determine how each student performed in relation to other students, compute percentile ranks and/or what other type of scores?

8. If the mean was selected as the average, what statistic should be computed to examine the variability in a set of scores?

9. There were two scores per research participant in which of the three examples ("Example 1," "Example 2," *or* "Example 3")?

10. To interpret a value of a Pearson r, which statistic should be computed?

Solution section:

> four (mindless)   nominal (is) p (are)   % (strength) n (very)   three (it)
>
> range (people)   standard scores (of)   Example 1 (wrong)   frequencies (crazy)
>
> experimental (beautiful)   mean (sign)   coeffecient of determination (memory)
>
> Example 3 (bad)   ratio (invent)   median (to)   standard deviation (a)
>
> ordinal (heaven)   number-right scores (kindness)   two (death)   control (often)
>
> interquartile range (clear)   Example 2 (weakness)   interval (the)

Worksheet 55 (Continued)

Write the answer to the riddle here, putting one word on each line: _____ _____ _____ _____

_____ _____ _____ _____

_____

Computer Repairs

GLASBERGEN

PHIL

"Your computer has high cholesterol and
an overdose of statistics homework."

Section 56: Inferential Statistics: Their Value in Research

As you know from Part B of this book (Sections 27 through 54), a wide variety of *inferential statistics* are used to interpret descriptive statistics in light of *sampling errors*, which are defined as "errors created by random sampling." Consider Example 1 to review one application of inferential statistics.

Example 1:
Members of the school board of a large school district asked researchers to estimate their high school seniors' knowledge of current events. Instead of administering a current events test to all 1,950 seniors in the district, the researchers administered it only to 300 of the seniors who were selected using simple random sampling (like drawing names out of a hat). On the 50-item current events multiple-choice test, the mean score was 35.0, which the researchers reported to the school board. The researchers cautioned the members of the board, however, to keep in mind that the mean of 35.0 was only an *estimate* of the average based on the simple random sample of 300 seniors. They also reported that a safer estimate of the seniors' knowledge was the **95% confidence interval** (see Sections 31 and 32 in this book) for the mean, which the researchers calculated to be 31.0 to 39.0 based on the performance of the sample of 300 seniors. Thus, board members were able to have 95% confidence that the true mean (which would have been obtained if the researchers had tested all seniors in the district) would have been between 31.0 and 39.0. Based on this information, the board members voted to hire consultants to advise the board on how to increase students' knowledge of current events.

The researchers in Example 1 reported the mean and standard deviation as the *descriptive statistics*, which describe only the seniors in the sample of 300.

They calculated the 95% confidence interval using inferential statistics. Note that testing a random sample of only 300 is much more efficient and cost-effective than testing all 1,950 seniors in the district. However, without inferential statistics (such as the limits of the 95% confidence interval for the mean), it would be meaningless to try to interpret the mean based on a sample because there would be no estimate of the extent to which random errors affected the results. Thus, Example 1 illustrates the first major value of inferential statistics: They help us interpret statistics that estimate population values (such as the mean of a population) based on studies of random samples.

The second major value of inferential statistics is a corollary of the first: Inferential statistics help us interpret differences in descriptive statistics among various samples drawn at random. Consider Example 2 to review this valuable contribution of inferential statistics.

Example 2:

A social science researcher conducted an experiment to explore the effectiveness of two programs for helping former employees of a large corporation who had been laid off due to a recession find employment with other companies. Program A stressed individual job placement counseling while Program B stressed group job placement counseling. The researcher selected a random sample of the laid-off employees and assigned them to Program A and assigned another random sample to Program B. After the programs were administered to the two samples, the researcher determined that 22% of those in Program A found new employment while only 20% of those in Program B found new employment. By conducting an inferential test of statistical significance, the researcher determined that the probability of the difference between 22% and 20% was not statistically significant. In light of this result, the researcher reported to the corporation's directors that the difference in the outcome between the two programs was statistically insignificant (i.e., an unreliable difference that could have been created by the random selection of the groups). On the basis of the researcher's experiment, the corporation's directors realized that the two treatments appeared to be about equal in their effectiveness. Given this information, they selected Program B for use in the future because the group counseling component of that program was less expensive than the individual counseling component of Program A.

As you probably recall, a test of statistical significance tests the **null hypothesis** (see Sections 33 and 34). In general, the null hypothesis states that any differences that are obtained when using random sampling are the result of sampling error. In Example 2, these errors would be those created by the random selection of participants for the two groups (i.e., the two groups might differ in ways that might affect the outcome of the experiment quite at random). The researcher found that the difference was not greater than what would be expected on the basis of random error alone. Hence, the result is insignificant.

In Example 2, the researcher tested for the significance of the difference between frequencies (and the corresponding percentages) using the inferential statistical test named the **chi square test** (see Sections 48 through 50). Now, consider Example 3, in which a different test of significance was used in a nonexperimental study.

Example 3:
A researcher wanted to know if there was a difference in job satisfaction between those who were hired before a new, less generous retirement program was instituted and those who were hired later. The researcher drew a random sample of each group and administered a standardized job satisfaction scale that yielded scores from 20 to 80. The researcher obtained a mean of 70.50 for the sample hired under the old retirement program and a mean of 62.75 for the sample of those hired later. A t test yielded a probability of $p <$

.001, meaning that it was highly unlikely (the likelihood was less than 1 in 1,000) that the difference between the two means was the result of random errors created by the random sampling. Because it was highly unlikely that the difference was due to sampling error, the researcher concluded that there was a statistically significant difference between the two means and, therefore, rejected the null hypothesis.

Notice that in Example 3, the *t test* (see Sections 37 through 40) helped in the interpretation of the descriptive statistics (i.e., the difference between the two means). Without an inferential test, the researcher could not determine whether the difference was greater than would be expected on the basis of chance alone (e.g., quite at random, the researcher might have selected more satisfied employees from those under the old retirement program).

In summary, inferential statistics help us interpret descriptive statistics in light of possible errors created by random sampling. Studying only a random sample is very often more efficient than studying whole populations, and inferential statistics help us make inferences about what populations are like based on random samples drawn from the populations.

Terms to Review Before Attempting Worksheet 56

sampling errors (see Sections 27 and 28), **95% confidence interval** (see Sections 31 and 32), **null hypothesis** (see Sections 33 and 34), **chi square test** (see Sections 48–50), *t test* (see Sections 37–40)

Worksheet 56: Inferential Statistics: Their Value in Research

> *Riddle*: What is the quickest way to double your money in Las Vegas?

DIRECTIONS: To find the answer to the riddle, write the answer to each question in the space immediately below it. The word in parentheses in the solution section next to the answer to the first question is the first word in the answer to the riddle, the word beside the answer to the second question is the second word, and so on.

1. According to the information in this section, is there a "wide" *or* "narrow" variety of inferential statistics?

2. *Inferential statistics* are used to interpret what other type of statistics?

3. Did the researchers in Example 1 administer the current events test to all 1,950 seniors in the school district?

4. In addition to the standard deviation, what other descriptive statistic was reported in Example 1?

5. The second major value of inferential statistics is that they help us interpret differences in descriptive statistics among various samples drawn at_____?

Worksheet 56 (Continued)

6. When a difference is unreliable, is it "statistically significant" *or* "insignificant"?

7. A *t* test was used in which Example (1, 2, *or* 3)?

8. What adjective is used before the word "hypothesis" to indicate that the hypothesis is the one that states that any differences obtained when using random sampling are the result of sampling error?

9. What is the name of the inferential statistical test used in Example 2 to determine whether the difference was statistically significant?

10. When an inferential test indicates that $p < .001$, this means that the likelihood is less than 1 in how many others?

11. What type of sampling was used in all three examples?

Worksheet 56 (Continued)

Solution section:

percentages (blackjack)   populations (climb)   descriptive (it)   100 (spend)

Example 2 (jumping)   insignificant (put)   errors (Nevada)   null (back)   no (in)

wide (fold)   significant (steal)   stratified (chips)   Example 1 (ladder)   1,000 (your)

random (and)   Example 3 (it)   mean (half)   inferential (were)   volunteers (bank)

efficient (casino)   chi square (in)   research (having)   random (pocket)

yes (being)   frequencies (broke)   10,000 (lose)   deviation (dealer)

Write the answer to the riddle here, putting one word on each line: _____ _____ _____ _____ _____ _____ _____ _____ _____ _____

STATISTICS 101

If at first you don't succeed, blame your computer.

GLASBERGEN

Notes:

Section 57: Limitations of Inferential Statistics: I

As you know from Part B of this book (Sections 27 through 54) and Section 56, *inferential statistics* are used to interpret descriptive statistics in light of *sampling errors*, which are defined as "errors created by random sampling" (see Sections 27 and 28). Specifically, inferential statistics are used to build *confidence intervals* (see Sections 31 and 32) as well as to interpret differences among descriptive statistics for various samples drawn at random (such as comparing the means of two samples drawn from two different populations).

As you probably recall, a test of statistical significance tests the *null hypothesis* (see Sections 33 and 34). The null hypothesis asserts that any differences obtained when using random sampling are the result of sampling errors created by the process of random sampling. Inferential statistics test this assertion. When they indicate that there is a low probability that the null hypothesis is true (such as $p < .05$, .01, or .001), it is conventional to reject the null hypothesis and conclude that it is unlikely that any difference being considered is the result of sampling errors.

Despite their great value in research, inferential statistics have two major limitations that are often insufficiently recognized. In this section, we will consider the first limitation: In the strictest sense, inferential statistics are valid only for interpreting statistics obtained from *random samples. Nonrandom samples* (such as a sample of people who happen to be in a shopping mall as representatives of all shoppers in a city *or* using just the students in one classroom as a sample of all students in a school because only one teacher agreed to allow the research to be conducted) are, by default, considered to be *biased samples*. In other words, if a sample is not drawn at random, it is presumed to be biased (see Sections 27 and 28 to review types of bias and various methods of random sampling). Consider Example 1 as background material for a more detailed discussion of this first limitation of inferential statistics.

Example 1:

A parent was interested in other parents' reactions to the possibility of requiring elementary school students in a large public school district to wear uniforms to school. The parent selected several nearby elementary schools and briefly interviewed some of the parents who were picking up their children at the end of the school day. In all, 50 parents were asked to rate the desirability of requiring uniforms on a scale from 10 (extremely desirable) to 1 (extremely undesirable). A mean of 7.25 was obtained for the 35 women parents and a mean of 6.75 was obtained for the 15 men parents who were interviewed. These descriptive statistics were presented at an open meeting of the school board as evidence that parents (on average) favor requiring students to wear uniforms, while noting that the women parents were more in favor than the men parents.

You probably recognize that the samples of women and men in Example 1 are biased. Perhaps the most important source of bias results from selecting only parents at "nearby" elementary schools within the district's total area. Thus, the sample is biased against parents whose children attend "far-away" elementary schools. The geographical bias (i.e., the bias against parents at "far-away" schools) is potentially very serious because neighborhoods tend to differ in many important respects such as socioeconomic status, political affiliation, cultural/ethnic/racial composition, immigrant status, and so on. Thus, what was learned from parents whose children attended nearby schools might be very different from what would be learned by studying a random sample from all the schools in the school district. Unfortunately, there are no generalizable mathematical tools or models (including no generalizable inferential statistics) to deal with specific biases (such as the bias against "far-away" schools).

You also may have noticed in Example 1 that the parent conducting the study interviewed only parents who were picking up their children from school. Parents who pick up their children might differ in a vast number of ways from those who do not. Those who do so might be more protective, have younger children, have children who live beyond walking distance from the school, be unemployed (and, thus, have time to pick up their children), and so on. Once again, there are no generalizable inferential statistical methods for interpreting errors due to biases. Inferential methods are designed for interpreting descriptive statistics that may have been influenced by random errors only.

It is important to note that inferential statistics can be calculated from any set of values (such as scores from 1 to 10 in Example 1) whether or not the scores were obtained from a random sample. Thus, it is *mathematically possible* to calculate confidence intervals for the two means in Example 1 as well as to perform the calculations necessary to conduct a *t* test of the significance of the difference between the means for women and men parents. However, these inferential statistics would be of limited value (from a strict point of view) because they would have been misapplied to descriptive statistics obtained from clearly biased samples of women and men parents.

Although researchers know the value of using randomly drawn samples, as a practical matter they often have great difficulty in obtaining them when studying humans as participants in research. This is true for two reasons. First, sometimes it is impossible to identify all members of a population, which makes it impossible to draw a random sample of all of them. (Note, for instance, that we cannot put all the names in a hat in order to draw a random sample if we do not know all the names of the members of the population.) For example, it is not possible to identify all individuals with certain diseases (partly because many people with the disease may be undiagnosed or because of privacy issues). Likewise, it is not possible to identify all successful (never arrested) burglars in a state, all the homeless in a city, all the married couples in abusive relationships, and so on. The second reason why it is often difficult or impossible to obtain random samples is that in most settings, researchers must rely on voluntary participation of subjects. For instance, a research psychologist might draw at random a sample of all freshmen on a college campus (a known population), but be required by the college to use in his or her research only those in the sample who agree in writing to participate in the research. Those who do not agree and, therefore, do not participate in the research, create a bias in the remaining sample (i.e., those who participate constitute a sample biased in favor of the type of people who tend to agree to participate in psychological research).

Despite the difficulties in obtaining random samples from important populations of human subjects, researchers proceed with their investigations in order to obtain at least limited information on pressing problems in fields in which it is often difficult to obtain random samples such as medicine, sociology, psychology, political science, and education. Furthermore, it has become conventional to apply inferential statistics in studies in which biased samples are used and to report these statistics in published research reports. When this is done, inferential statistics provide us with information on a "what if" scenario. In other words, inferential statistics are being used to explore these questions: "What if the sample were random instead of biased?" "What confidence intervals and levels of significance would be obtained under this 'what if' condition?"

It is important to note that when inferential statistics are applied to results obtained with biased samples, researchers and consumers of research should use great caution in interpreting the results because in the strictest sense, inferential statistics should be limited to analysis of data obtained from random samples. Nevertheless, application of inferential statistics provides us with hints as to what might be obtained under more optimal circumstances. In addition, they provide a common standard for comparing the results from one study to the next (e.g., to identify which studies had low values of p and which did not) even if the studies were conducted with biased samples.

Because of the limitation considered in this topic, the results obtained with inferential statistics should be taken with a grain of salt whenever biased samples are used. When a sample is very clearly biased, such as in Example 1, researchers are advised to warn their readers of the limitation. In many cases, it is also a good idea to label studies with this weakness as *pilot studies* in either the title and/or introduction and conclusion of the research reports.

Note that the limitation we have considered in this section is *not* due to some mathematical flaw in the development of inferential statistics. Instead, the limitation results from the fact that inferential statistics were designed for interpreting the results that have errors due to random sampling. In the same sense that even the very best traditional automobile is of limited value for off-road driving, inferential statistics are of limited value for interpreting the results obtained from biased samples.

In the next section, you will learn about another important limitation of inferential statistics.

Terms to Review Before Attempting Worksheet 57

inferential statistics (see Section 56), **sampling errors** (see Sections 27 and 28), **confidence interval** (see Sections 31 and 32), **null hypothesis** (see Sections 33 and 34), **biased samples** (see Sections 27 and 28)

"You are getting sleepier and sleepier. When I count to three, you will be able to answer all my statistics homework questions correctly."

Worksheet 57: Limitations of Inferential Statistics: I

> *Riddle*: What does the wacky sign outside the
> muffler shop say?

DIRECTIONS: To find the answer to the riddle, write the answer to each question in the space immediately below it. The word in parentheses in the solution section next to the answer to the first question is the first word in the answer to the riddle, the word beside the answer to the second question is the second word, and so on.

1. What adjective is used before the word "errors" to indicate errors that are created by random sampling?

2. Inferential statistics can be used to build confidence_____.

3. Is the null hypothesis rejected when the probability determined with a significance test is "high" *or* when it is "low"?

4. If a sample is not drawn at random, it is presumed to be what?

5. Should the samples of men and women parents in Example 1 be presumed to be *unbiased*?

6. If a sample is biased, is it mathematically possible to calculate inferential statistics?

Worksheet 57 (Continued)

7. What adjective might be put before the word "study" in either titles and/or introductions and conclusions of research reports to warn consumers of research to use caution in making inferences about populations from studies in which biased samples are used?

Solution section:

inferential (cars)   intervals (appointment)   biased (we)   descriptive (free)

unbiased (used)   samples (noise)   calculate (cost)   high (sound)   default (ticket)

pilot (coming)   probability (shopping)   large (mechanic)   sampling (no)

confidence (driving)   hypothesis (money)   published (moving)   no (hear)

low (necessary)   confidence (charge)   hypothesis (loud)

errors (check)   geographical (dangerous)   yes (you)

Write the answer to the riddle here, putting one word on each line: _____   _____   _____.   _____ _____   _____   _____.

Notes:

Section 58: Limitations of Inferential Statistics: II

From the previous section, you know that inferential statistics are of limited value in interpreting descriptive statistics obtained by studying biased samples. Even if unbiased (i.e., random) samples are used, there is an additional limitation: Inferential statistical tests indicate only whether differences are *reliable*; they do *not* indicate whether differences are large. Those who mistakenly believe that all significant differences are large will often be misled.

To understand this limitation, first consider the difference between a "reliable difference" and a "large difference." When we say that we have identified a "reliable" difference, we are referring to "consistency," not whether it is large or small. Example 1 illustrates that a small difference can be reliable.

Example 1:
Jackson arrives at work each morning just in the nick of time, clocking in before his pay is docked. He would prefer to arrive a few minutes early and not have to rush. However, he is a single parent of an elementary school child, and he waits until his daughter is on the school bus before leaving for work. This leaves him just enough time to rush to work and clock in just on time. His behavior is *reliable* because it is consistent from day to day. On the other hand, Sheila arrives about five minutes early each day because the public bus that she takes arrives at that time. Sheila's behavior is also highly *reliable* because it is consistent from day to day.

Under most circumstances, a five-minute difference such as the one between Jackson and Sheila in Example 1 would be considered "small." In the example, the difference is not only small but it is also unimportant since both employees arrive at work on time. Despite its smallness and unimportance, over a long enough period of observations, anyone would become convinced that this is a real difference and not one due to chance factors alone. In the same fashion, over a large enough sample, significance tests can identify reliable (i.e., statistically significant) differences even if they are very small.

To better understand why small differences can be statistically significant differences, consider the three basic factors that underlie the *t test*. As you know from Section 37, they are:

1. The size of the sample: *The larger the sample, the more likely that the difference will be found to be statistically significant.*

2. The difference between two means: *The larger the difference, the more likely that the difference will be found to be statistically significant.*

3. The *variance* among the subjects: *The smaller the variance, the more likely that the difference will be found to be statistically significant.*

Thus, while a result with a larger difference is more likely to be statistically significant (see point 2 immediately above), the other two factors (sample size and variance) also contribute in determining whether the difference between two means is significant. In fact, if sample size is very large and variance is very small in a study that has a small difference between means, the sample size and variance can "overwhelm" the small difference and create a statistically significant result. It is important to note that this is not a flaw of significance testing. Rather, it is a limitation because significance testing was designed to help us identify only *reliable* (i.e., consistent) differences—not necessarily large differences. If this is confusing, consider Example 1 again and use it as an analogy that illustrates that small differences can be reliable, and remember that significance tests were designed to test for reliability.

Individuals who do not understand this limitation of significance testing often make correct but potentially misleading statements such as these:

1. Program A should be adopted because studies show that it is, on average, significantly superior to Program B.

2. The new college admissions test is valid because the scores on it are significantly correlated with freshman GPA; the test is a statistically significant predictor of college grades.

3. The new drug should be approved by the FDA because significantly more individuals in the experimental group than in the control group reported relief from their symptoms.

All three statements are potentially misleading because they refer only to significance without describing the size of the difference. For the first statement, we would want to know on what scale the outcome was measured and the values of the means (and standard deviations). For the second statement, we would want to know the value of the correlation coefficient. For the third statement, we would want to know the frequencies and percentages of cases in each group that reported relief from symptoms. In short, we cannot determine whether the differences are large without examining the descriptive statistics on which the inferential significance tests were performed. Because of this, it is traditional to report descriptive statistics first and then report inferential statistics in the results sections of research reports.

Terms to Review Before Attempting Worksheet 58

reliable, *t* **test** (see Section 37), **variance** (see Sections 12 and 13)

Worksheet 58: Limitations of Inferential Statistics: II

Riddle: What is the newest modern law of physics?

DIRECTIONS: To find the answer to the riddle, write the answer to each question in the space immediately below it. The word in parentheses in the solution section next to the answer to the first question is the first word in the answer to the riddle, the word beside the answer to the second question is the second word, and so on.

1. "The only purpose of inferential statistics is to indicate how large a diference is." Is this statement "true" *or* "false"?

2. When we say we have identified a "reliable" difference, we are referring to what?

3. Is the difference in Example 1 described as being "large" *or* is it described as being "small"?

4. Is it possible for significance tests to identify reliable differences even if they are very small?

5. How many basic factors underlie the *t* test?

6. "The larger the sample, the more likely that the difference will be found to be statistically significant." Is this statement "true" *or* "false"?

Worksheet 58 (Continued)

7. "The larger the variance, the more likely that the difference will be found to be statistically significant." Is this statement "true" *or* "false"?

8. Are studies with larger samples more likely to have significant differences than studies with smaller samples?

9. Is the limitation discussed in this section a "flaw"?

10. For the second numbered statement near the end of this section (statement 2), we would want to know the value of the correlation_____.

11. In short, we cannot determine whether differences are large without examining what type of statistics (other than inferential statistics)?

Worksheet 58 (Continued)

Solution section:

two (physical)   true (at)   one (measures)   no (speed)   significant (sound)

three (mail)   hypothesis (being)   false (bills)   significance (travel)

coefficient (of)   five (assuming)   yes (the)   research (stamps)   sample (of)

small (through)   percentages (sometimes)   predictors (come)

descriptive (checks)   sampling (generalize)   yes (the)   variation (though)

deviation (possibility)   consistency (travel)   correlational (people)

Write the answer to the riddle here, putting one word on each line: _____ _____ _____ _____ _____ _____ _____ _____ _____ _____ _____

Section 59: Statistical Versus Practical Significance

In the previous section, you learned that statistically significant differences are not necessarily large differences. Likewise, a statistically significant correlation coefficient is not necessarily large.[1] Because of this, it is important to carefully consider the descriptive statistics before drawing conclusions and considering the implications of statistically significant results. This is illustrated in the next two examples.

Example 1:

A college admissions officer developed an experimental admissions test and administered it to a random sample of the incoming freshmen in September. At the end of their freshman year, he correlated the admission test scores with freshman-year GPA and obtained a *Pearson r* of 0.20 (see Sections 21 and 22). For the sample size used, the value of 0.20 was found to be statistically significant at the .05 probability level. To interpret the results further, the researcher calculated the *coefficient of determination* (a descriptive statistic described in Section 23), which equals 0.04 in this case. Multiplying 0.04 by 100%, the researcher learned that the amount of variation in GPA accounted for by the experimental test was only 4.0%. The researcher subjectively concluded that the experimental test is not an important predictor of freshmen GPA, even though the underlying correlation coefficient is statistically significant.[2]

---

[1]When a test of statistical significance is conducted on a correlation coefficient, the *difference* between the value obtained with a random sample of subjects is compared with a hypothetical "null value" of 0.00. If the value observed by the researcher is determined to be statistically significant, we know that the difference between the observed value and a value of 0.00 is reliable (i.e., unlikely to be due to sampling error).

[2]The researcher could have assessed the contribution of the experimental test in combination with other predictors by using *multiple correlation* (see Section 24).

Note that the decision on the importance of the result in Example 1 also can be made in light of the context of what is already known about the effectiveness of other predictors of college grades. For instance, high school grades often account for about 25% of the variance in college freshmen grades. In this context, the 4% discussed in Example 1 does not seem especially important.

Example 2:

An educational researcher administered a standardized math test to a random sample of third graders in a large urban school district. The researcher also administered the test to a random sample of students in the suburban school districts that surround the urban school district. The difference between the two means was found to be statistically significant at the .01 probability level. Before reaching conclusions, the researcher reexamined the means, which were 45.75 for the urban sample and 55.75 for the suburban sample. Because the standardized test yields **T scores** that can range from 20.00 to 80.00 (see Sections 16 through 18), the researcher reported that the 10-point difference between the two means was educationally important because it was 1/6 of the total possible difference that could be obtained (i.e., the 10-point difference was 1/6 of the maximum possible difference of 60 points from 20.00 to 80.00), *and* it was a reliable difference (i.e., statistically significant). This decision on educational importance was made subjectively. However, it had a statistical basis (i.e., the researcher had considered both the magnitude of the difference indicated by the descriptive statistics as well as the result of the significance test).

In Example 2, the data were obtained with a standardized test that yields easily interpreted T scores. In much research, nonstandardized tests that have been custom-made for the particular research objectives are frequently used. When this is the case, comparing the possible maximum range of the difference with the obtained difference between means is helpful in reaching conclusions. For instance, if overall attitude toward the homeless is measured with a statement that respondents rate on a scale from 1 to 10, you know that the maximum possible range is 9 (i.e., $10 - 1 = 9$). If two groups are compared and a statistically significant difference between their two means is 3 points on such a scale, then we know that this is a third of the maximum possible difference. While 3 points might not be considered important when using a test with a maximum possible difference of 600 points, such as *CEEB scores* that range from 200 to 800 (see Section 17), when using a measure with a possible range of only 9 points, the 3-point difference might be regarded as very important.

Up to this point, we have been considering the "importance" of differences. In applied research, the question of importance is often discussed in terms of *practical significance*. A statistical result is of practical significance when it has direct implications for professionals in applied fields such as clinical psychology, social work, and education.

It is important to note that statistical significance is *not* necessary for a result to be of practical significance. For instance, suppose the administrators of a social work department were considering using a new computerized system for tracking certain types of clients. If an experiment revealed that the difference between the outcomes obtained by using the new system (which would have considerable new costs associated with it) and the outcomes obtained by using the tracking system already in use were not statistically significant, the following conclusion, which is of considerable practical importance, might be reached: Do *not* purchase and implement the new system.

Determining practical significance requires consideration of a number of issues in addition to direct statistical considerations. We will briefly consider the major ones here. The first issue is *cost in relation to benefit*. This is often thought of in terms of the *cost for each unit of benefit*. For instance, if an experiment showed that a new computerized math program was superior to the current program on the average by 6 points and that the difference was statistically significant, we could ask how much it would cost for each of the 6 points of difference. (In this case, the cost per unit can be determined by calculating how much more the new program would cost than the old program and dividing the difference by 6.) If the cost per unit is subjectively judged to be excessive, we would conclude that the new program is not of practical significance.

The second major issue in determining practical significance is whether there are *side effects*. Researchers who study the effects of prescription and over-the-counter drugs almost always design their studies so that they will detect undesirable side effects. Likewise, researchers in other fields often formally study side effects. For instance, research on mathematics achievement often includes measures of attitudes toward math because a program that produces superior math achievement but causes a deterioration of attitudes toward math might be of limited practical value.

The third issue is *acceptability* to clients, students, parents, and other stakeholders. For instance, if statistics indicate that there are benefits from requiring public school students to wear uniforms but the parents in a particular school district are strongly opposed to requiring uniforms, a policy requiring them might not be politically acceptable to the school board.

Practical significance is also influenced by *legal and ethical issues*. Programs, treatments, and other procedures that are found to be statistically superior but are illegal or unethical obviously should be avoided.

In short, descriptive statistics summarize and organize the results of a study so that they can be understood more fully and communicated concisely. Inferential statistics tell us whether any differences are reliable in light of the possibility that the differences were created by random error. Determining practical significance is the last step in the research process, and it is based on sound judgment that goes beyond mathematical reasoning.

Terms to Review Before Attempting Worksheet 59

Pearson *r* (see Sections 21 and 22), **coefficient of determination** (see Section 23), **multiple correlation** (see Section 24), ***T* scores** (see Sections 16–18), ***CEEB* scores** (see Section 17), **practical significance, cost in relation to benefit, side effects, acceptability, legal and ethical issues**

Worksheet 59: Statistical Versus Practical Significance

> *Riddle*: What is the new motto of the Internal Revenue Service?

DIRECTIONS: To find the answer to the riddle, write the answer to each question in the space immediately below it. The word in parentheses in the solution section next to the answer to the first question is the first word in the answer to the riddle, the word beside the answer to the second question is the second word, and so on.

1. Was the correlation coefficient in Example 1 statistically significant?

2. Did the researcher in Example 1 decide that the correlation was important?

3. In Example 2, was the decision on educational importance made "subjectively" *or* "objectively"?

4. "The decision on educational importance in Example 2 has a statistical basis." Is this statement true *or* false?

5. "Standardized tests are almost always used in research." Is this statement true *or* false?

6. A difference is of practical significance when it has what?

Worksheet 59 (Continued)

7. What is the "second major issue" in determining practical significance?

8. What is the third issue in determining practical significance?

9. Practical significance is also influenced by *legal* and what other kinds of issues?

10. Determining practical significance is which step in the research process?

Solution section:

true (it)   test (owe)   subjectively (what)   *T* (if)   false (takes)   *CEEB* (almost)

size (only)   direct implications (to)   significant (taxes)   no (got)   benefit (check)

unit (refund)   yes (we've)   null (government)   ethical (you've)   research (money)

cost (dollars)   probability (audit)   hypothesis (welfare)   side effects (get)

hypothesis (exempt)   sampling (claim)   acceptability (what)   last (got)

objective (going)   error (strong)   practical (in)   practicality (dollars)

Write the answer to the riddle here, putting one word on each line: _____ _____ _____ _____

_____ _____ _____ _____ _____

_____

SUPPLEMENT: Basic Math Review

The sections in this supplement show you how to perform basic arithmetic operations that are required in statistics. Even if you will be using a calculator or computer, you need to understand them so you can check the reasonableness of your answers and catch errors. The *answers* to many of the exercise problems at the end of each section are given at the end of this supplement.

Section A Order of Operations

In statistics, you will be using the four basic operations: addition, subtraction, multiplication, and division. Keep in mind that:

(3)(7) means "multiply 3 times 7."

6/2 means "6 divided by 2."

The result of addition is known as the *sum*.

The result of multiplication is known as the *product*.

The result of subtraction is known as the *difference*.

The result of division is known as the *quotient*.

✔ **Rule One:** When there are parentheses, perform the operations inside the parentheses first. Here are three examples:

$(4)(2 + 1) = ?$

Thus, $(4)(3) = 12$

$(4 - 3)/(5 - 4) = ?$

Thus, $1/1 = 1$

$(9)(7 + 3 - 2) = ?$

Thus, $(9)(10 - 2) = ?$

and $(9)(8) = 72$

✔ **Rule Two:** Unless parentheses indicate otherwise, multiply and divide before adding and subtracting.

In these examples, you must multiply before adding:

$$5 + (3)(2) = ?$$
Thus, $5 + 6 = 11$

$$8 + (5)(10) + 1 = ?$$
Thus, $8 + 50 + 1 = 59$

In these examples, you must divide before subtracting:

$$6 - 2/2 = ?$$
Thus, $6 - 1 = 5$

$$20 - 36/6 - 5 = ?$$
Thus, $20 - 6 - 5 = 9$

In these examples, you must multiply and divide before adding and subtracting:

$$(5)(2) - 1 + 4/2 = ?$$
Thus, $10 - 1 + 2 = 11$

$$11 + (10)(3) - 4/2 = ?$$
Thus, $11 + 30 - 2 = 39$

✔ **Rule Three:** If there are both parentheses and brackets, first solve within the parentheses, then within the brackets, and then perform any remaining operations.

Study the following examples.

$$[(2)(2) - (5 - 3)][7 - 1] = ?$$
Thus, $[4 - 2][6] = ?$
and $[2][6] = 12$

$$[(25)(10 - 5)]/5 = ?$$
Thus, $[(25)(5)]/5 = ?$
and $125/5 = 25$

Exercise for Section A

(Note: The answers to items 1–12 are given at the end of this supplement.)

1. $(10)(3 + 5) = ?$
2. $(5 + 4)/(2 + 1) = ?$
3. $(6 - 5 + 2)(5) = ?$
4. $8 + (5)(4) = ?$
5. $(10)(11) - 1 = ?$
6. $5 + 12/4 = ?$
7. $10 + (2)(5) - 5 = ?$
8. $25 - (9)(2) + 3 = ?$
9. $[(4 + 7)(3 - 1)][8 - 3] = ?$
10. $[(3 + 5) + (1)(2)]/2 = ?$
11. The result of multiplication is known as the
    A. product. B. quotient. C. sum. D. difference.
12. The result of addition is known as the
    A. product. B. quotient. C. sum. D. difference.
13. $(4 + 6)(11) = ?$
14. $(7 - 1 + 2)(4) = ?$
15. $20/(5 + 5) = ?$
16. $9 + 8/2 = ?$
17. $(12)(12) - 3 = ?$
18. $9 + (4)(8) = ?$
19. $15/3 + 5 - 6/2 = ?$

20. $19 + (2)(6) - 4 = ?$

21. $[(5 - 3) + (2)(3)]/8 = ?$

22. $[(9 + 3) - (1)(3)][6 - 2] = ?$

23. The result of division is known as the

   A. product. B. quotient. C. sum. D. difference.

24. The result of subtraction is known as the

   A. product. B. quotient. C. sum. D. difference.

Section B   Squares and Square Roots

In statistics, you will be squaring many numbers. To square a number, multiply it by itself as in this example:

$$4^2 = 4 \times 4 = 16$$

Note that in the example, the 4 is the *base* and the ^{2} is the *exponent*.

CALCULATOR HINT: To square a number using a calculator, you only have to enter the number once. For example, on your calculator:

   1. Press 4.

   2. Press the times sign ($\times$).

   3. Press the equal sign (=).

   4. You should see 16, which is the square, displayed.

Taking a square root is the opposite of squaring. Thus, the square root of 16 is 4. To take a square root on a calculator, just enter the number and then press the square root sign ($\sqrt{}$). The formal name of the square root sign is the *radical sign*.

   Notice that:

$$\sqrt{16} \text{ is read } \textit{the square root of 16}, \text{ which is 4.}$$

When a square root sign (i.e., radical sign) appears in a formula, it has the same effect as parentheses on the order of operations. That is, everything under a radical sign must be solved and the square root taken before performing any operations on its value. For example, the following tells us to multiply 4 times the sum under the radical sign.

$$4\sqrt{20+5} = ?$$

Because of the radical sign, you must sum 20 and 5 and take the square root of the sum before multiplying by 4:

Thus, $4\sqrt{25} = ?$
and $(4)(5) = 20$

Notice that 4^2 is read *four squared*, while $\sqrt{4}$ is read *the square root of 4*.

Exercise for Section B

(Note: The answers to items 1–10 are given at the end of this supplement.)

1. 14^2 is read

   A. 14 squared. B. the square root of 14. C. double 14.

2. The "14" in Question 1 is known as the

   A. base. B. exponent. C. radical sign.

3. $\sqrt{9}$ is read as

   A. 9 squared. B. the square root of 9. C. half of 9.

4. The $\sqrt{}$ sign is known as the

   A. base. B. exponent. C. radical sign.

5. What is the square root of 144?

6. What is the square of 100?

7. $\sqrt{225} = ?$

8. $18^2 = ?$

9. $6\sqrt{100-19} = ?$

10. $\sqrt{324} + 2(11-9) = ?$

11. $\sqrt{49}$ is read

   A. 49 squared. B. the square root of 49. C. double 49.

12. In 9^2, the "2" is known as the

    A. base. B. exponent. C. radical sign.

13. 67^2 is read as

    A. 67 squared. B. half of 67. C. the square root of 67.

14. The "2" in Question 13 is known as the

    A. base. B. exponent. C. radical sign.

15. What is the square root of 81?

16. What is the square of 60?

17. $\sqrt{169} = ?$

18. $51^2 = ?$

19. $4\sqrt{55 - 6} = ?$

20. $\sqrt{250 + 6} / (3 + 5) = ?$

Section C Negatives

We frequently use negative numbers in statistics. Because we mainly use positive numbers in everyday situations, your knowledge of negative numbers and how to operate on them may be rusty, so we will start with the basics. First, you may recall that a number line looks like this:

    –5 –4 –3 –2 –1 0 +1 +2 +3 +4 +5

Negative numbers are to the left of zero.

When a number is shown without a sign, it is understood to be positive. Thus, "3" is understood to be "+3." When a number is negative, its sign is always shown.

When working with negatives, follow these rules:

✔ **Rule One:** When you multiply or divide numbers with different (unlike) signs, the result is negative.

Specifically, when you multiply a positive number by a negative number, the result (product) is negative. Thus:

$$(5)(-2) = -10 \quad \text{and} \quad (-2)(5) = -10$$

When you divide using a positive and a negative number, the result (quotient) is negative. Thus:

$$10/-2 = -5 \quad \text{and} \quad -10/2 = -5$$

✔ **Rule Two:** When you multiply or divide numbers with the same sign, the result is positive, as in these examples:

$$-50/-5 = 10 \quad \text{and} \quad 50/10 = 10$$
$$(-3)(-9) = 27 \quad \text{and} \quad (3)(9) = 27$$

✔ **Rule Three:** When you add a set of numbers, all of which are negative, the result (sum) is negative, as in these examples:

$$-1 + -4 + -3 = -8$$
$$-10 + -11 = -21$$

✔ **Rule Four:** When you add a positive number and a negative number, temporarily ignore the signs and *subtract* the smaller from the larger. Then assign to the sum the sign of the larger number. Study this example:

$$3 + -7 = ?$$

First, ignoring signs, $7 - 3 = 4$. Since 7 is larger than 3 and since the 7 was originally negative, the sum is negative. Thus, the answer is -4. It helps to understand this example if you refer to the number line at the beginning of this section. If you start at $+3$ and add to that 7 points in the negative direction (moving left from $+3$ seven points), you come to -4, which is the answer.

Here is another example:

$$15 + -3 = ?$$

First, subtract the smaller from the larger: $15 - 3 = 12$. The answer is a positive 12 because the larger number is positive.

✔ **Rule Five:** If there are some positive and some negative numbers, all of which are to be summed, first add all the positive numbers to get their sum. Then add all the negative numbers using Rule Three to get their sum. Then add the two sums using Rule Four. For example:

$4 + -2 + 5 + 6 + 6 + -1 + 7 + -3 = ?$

First, sum the positive numbers:

$4 + 5 + 6 + 6 + 7 = 28$

Then sum the negative numbers using Rule Three:

$-2 + -1 + -3 = -6$

Then sum the two sums using Rule Four:

$28 - 6 = 22$

The answer is +22 because the larger number is positive.

✔ **Rule Six:** When you subtract a negative from a negative, temporarily ignore the signs and *subtract*. The result is negative. For example:

$-10 - -2 = ?$

First, $10 - 2 = 8$.

Assigning a negative sign, the answer is –8.

Note that when we ignore the sign of a number, we are using its *absolute value*—its value without regard to sign. For example, +8 and –8 have the same absolute value; if you take away their signs, they are the same.

✔ **Rule Seven:** When you subtract a negative from a positive, the negative number becomes a positive. *Add* the two numbers to get the difference. For example:

$5 - -4 = ?$

Thus, $5 + 4 = 9$.

At first, this rule may be confusing. It helps to think about it as though it were a double negative (for example, – –4) in English. If someone says, "We are not (negative) sure that we are not (negative) going," they are making the positive statement that they, in fact, might be going.

Another way to look at Rule Seven is to consider the following number line. It illustrates that the difference between +5 and –4 is 9. (Remember that when you subtract, you are getting a difference.)

$$-4 \quad -3 \quad -2 \quad -1 \quad 0 \quad +1 \quad +2 \quad +3 \quad +4 \quad +5$$

Count the *spaces* between +5 and –4; there are nine spaces. Thus, the difference between +5 and –4 is 9.

✔ **Rule Eight:** When you subtract a positive from a negative, temporarily ignore the signs and *add* the numbers. The answer is negative. For example:

$-8 - 7 = ?$

First, $8 + 7 = 15$

The answer is –15.

CALCULATOR HINT: Most calculators have a ± button. Use this when you enter negative numbers. Here is an example:

To solve $(5)(-2) = ?$

1. Press 5.
2. Press × (the times sign).
3. Press 2 (enter it as a positive).
4. Press the ± button (this will make the 2 negative).
5. Press = (the equals sign).
6. The display should show –10, the answer.

Pressing the ± button informs the calculator that the most recently entered number is a negative number.

Exercise for Section C

(Note: The answers to items 1–17 are given at the end of this supplement.)

Suggestion: Before you solve these problems with a calculator, mentally estimate whether the answer to each will be positive or negative.

1. $(-8)(7) = ?$
2. $(4)(-50) = ?$
3. $49/-7 = ?$
4. $-121/11 = ?$
5. $-36/-6 = ?$
6. $(-12)(-10) = ?$
7. $-5 + -9 + -6 = ?$
8. $5 + -9 = ?$
9. $-12 + 8 = ?$
10. $-9 + 8 + -3 + 1 = ?$
11. $3 + -5 + 12 + -2 = ?$
12. $-29 - -14 = ?$
13. $-47 - -17 = ?$
14. $10 - -9 = ?$
15. $12 - -21 = ?$
16. $-20 - 18 = ?$
17. The *absolute value* of a number is its value

    A. when it is negative. B. without regard to its sign. C. with its sign.

18. $(-9)(12) = ?$
19. $(15)(-7) = ?$
20. $-66/11 = ?$
21. $169/-13 = ?$
22. $-12/-3 = ?$
23. $(-15)(-14) = ?$
24. $-8 + -7 + -9 = ?$
25. $10 + -13 = ?$
26. $-17 + 9 = ?$
27. $5 + -7 + -6 + 4 = ?$

28. $8 + -2 + 9 + -1 = ?$

29. $-30 - -29 = ?$

30. $-17 - -18 = ?$

31. $20 - -11 = ?$

32. $14 - -18 = ?$

33. $-40 - 11 = ?$

34. How is "$-18 - 14 = ?$" read?

    A. Minus 18 minus negative $14 = ?$ B. Negative 18 minus negative $14 = ?$

    C. Negative 18 minus positive $14 = ?$

Section D   Decimals and Rounding

As you probably recall:

    3.2 is *three and two-tenths*.

    3.02 is *three and two-hundredths*.

    3.002 is *three and two-thousandths*.

It is customary in statistics to report answers to the hundredths' or thousandths' place. In *informal discussions*, we sometimes refer to 3.02 as *three point zero two* and 3.002 as *three point zero zero two*.

Sometimes a number is less than one, such as:

    0.56, which is *zero and 56 hundredths*

    .56 is also *zero and 56 hundredths*; the zero is implied.

It is a good idea to show the zero, however, because it helps to call your readers' attention to the decimal point and the fact that the value is less than one.

Modern calculating devices have eliminated the need to keep track of decimal places when computing. However, these devises have not eliminated errors in entering the data, which leads to incorrect answers. Hence, the following calculator hint.

CALCULATOR HINT: Make mental estimates by first rounding to whole numbers before using your calculator. For example, to divide 9.362 by 3.190, mentally round 9.362 to 9 and 3.190 to 3. Since 9/3 = 3, the answer your calculator gives you to the problem should be close to 3. In fact, the answer is 2.935. If your calculator does not show an answer close to 3, you will know that you entered one or more wrong numbers or entered one or more decimal places at the wrong points.

Frequently, when you divide using a calculator, you will obtain a long string of decimal places such as:

$$7.7/3 = 2.566666666666$$

You will want to round this. To the nearest hundredth, it rounds to 2.57; to the nearest thousandth, it rounds to 2.567.

Review this terminology:

10.366 = 10.37 is an example of *rounding up*.

10.361 = 10.36 is an example of *rounding down*.

A special rule for rounding off five that is often used in statistics is:

✔ Round all numbers ending in five to the nearest even number.

Another way to state the same rule is:

✔ When rounding off the five in a number ending in five, round down if the preceding number is even and round up if the preceding number is odd.

Here are two examples of the application of the special rule for five:

10.335 = 10.34 (round up).

10.345 = 10.34 (round down).

By using this special rule, we will, in the long run, round up about half the time and down about half the time when a number ends in five. This is a compromise since there is no mathematical solution regarding how to round off a five. Remember that 5 is exactly halfway between 0 and 9.

If you are not comfortable with this special rule, study these examples:

113.485 = 113.48 (round down because the 8 is even).

27.65 = 27.6 (round down because the 6 is even).

9.1135 = 9.114 (round up because the 3 is odd).

896.875 = 896.88 (round up because the 7 is odd).

Notice that in all cases, after rounding off the five in a number that ends with five, the rounded numbers always end in an even number.

It is very important to notice that the special rule described above is *only for numbers that end in 5*. If a number ends in more than 5, round up. For example:

2.1659 = 2.17 when rounded to the nearest hundredth.

9.248 = 9.25 when rounded to the nearest hundredth.

If a number ends in less than 5, round down. For example:

1.454 = 1.45 when rounded to the nearest hundredth.

2.332 = 2.33 when rounded to the nearest hundredth.

Here is a rule for reporting answers: If you have zeros in the tenths' and hundredths' place, keep them if you are asked to report an answer to the nearest hundredth. For example, "35.004" should be rounded to "35.00" and reported as "35.00" and *not* as "35." By keeping the two zeros, you are showing that your answer is accurate to the nearest hundredth. A simple "35" may be a rounded result of some other value such as "35.23."

There are many multistep problems in statistics. That is, after you perform one computation, you will use the answer in the next computation; the second answer will be used in the third computation, and so on. At the end of each step, to how many places should you round your answer? Here are two general rules to guide you:

✔ The more decimal places you retain at the end of each step, the more accurate your answer will be.

✔ At the end of each step in a problem, retain at least one more decimal place than you will be reporting in the answer to the final step. For example, if you plan to report the answer to the final step in a problem to two decimal places, keep at least three in all steps leading to that answer. Then round the final answer to two decimal places.

If you keep a different number of decimal places at the end of each step in a problem than your instructor and classmates keep, you may obtain a slightly different answer. Slight differences in the hundredths' place are usually not of concern for all practical purposes—as long as they are attributable to rounding and not to an error.

Exercise for Section D

(Note: The answers to items 1–12 are given at the end of this supplement.)

1. "5.24 is read as *five and twenty-four tenths*." TRUE or FALSE?

2. "16.1 is read as *sixteen and one-tenth*." TRUE or FALSE?

3. "It is customary in statistics to report an answer to the nearest whole number." TRUE or FALSE?

4. "When rounded to a whole number, 5.189 becomes 5." TRUE or FALSE?

5. "The answer to 100.0812/9.8746 should be close to 10." TRUE or FALSE?

6. "The answer to (19.953)(5.162) should be close to 20." TRUE or FALSE?

7. "When rounded to the nearest hundredth, 6.4734 becomes 6.473." TRUE or FALSE?

8. "When rounded to the nearest thousandth, 7.5766 becomes 7.577." TRUE or FALSE?

9. "When rounded to the nearest tenth, 15.657 becomes 15.7." TRUE or FALSE?

10. "If the special rule is applied, 10.25 becomes 10.3 when rounded to the nearest tenth." TRUE or FALSE?

11. "If the special rule is applied, 111.665 becomes 111.66 when rounded to the nearest hundredth." TRUE or FALSE?

12. "If the special rule is applied, 15.5555 becomes 15.556 when rounded to the nearest thousandth." TRUE or FALSE?

13. "6.12 is read as *six and twelve-hundredths*." TRUE or FALSE?

14. "7.101 is read as *seven and one hundred and one thousandths*." TRUE or FALSE?

15. "It is customary in statistics to report answers to the nearest hundredth or thousandth." TRUE or FALSE?

16. "When rounded to a whole number, 3.119 becomes 4." TRUE or FALSE?

17. "The answer to 10.2 plus 40.18129 should be close to 50." TRUE or FALSE?

18. "The answer to 24.98762/4.823423 should be close to 5." TRUE or FALSE?

19. "When rounded to the nearest tenth, 8.129 becomes 8.2." TRUE or FALSE?

20. "When rounded to the nearest hundredth, 121.564 becomes 121." TRUE or FALSE?

21. "When rounded to the nearest thousandth, 14.47451 becomes 14.475." TRUE or FALSE?

22. "If the special rule is applied, 5.545 becomes 5.54 when rounded to the nearest hundredth." TRUE or FALSE?

23. "If the special rule is applied, 19.445 becomes 19.45 when rounded to the nearest hundredth." TRUE or FALSE?

24. "If the special rule is applied, 12.2535 becomes 12.253 when rounded to the nearest thousandth." TRUE or FALSE?

Section E Fractions

Many formulas in statistics contain fractions. In order to understand these formulas, it is essential that you know the basics about fractions, which are illustrated here by example:

In 1/6, 1 is the *numerator* and 6 is the *denominator*.

1/6 may be represented as follows. The whole consists of six equal parts,

and one of them (1/6) is shaded:

As the numerator increases, the value of the fraction increases. For example,

2/6 is greater than 1/6. Here, 2/6 of the area is shaded:

As the denominator increases, the value of the fraction decreases. For exam-

ple, 1/12 is less than 1/6. This represents 1/12:

Notice that 1/12 is half the size of 1/6.

When performing calculations involving fractions, follow these rules:

✔ **Rule One:** Use your calculator to convert fractions to their decimal equivalents and solve problems using the decimal equivalents. Suppose you have to multiply 29 by 1/4:

To solve $(29)(1/4) = ?$
First, convert 1/4 to its decimal equivalent by dividing 1 by 4, which equals 0.25.
Then multiply on your calculator $(29)(0.25) = 7.25$.

✔ **Rule Two:** If a formula specifies operations in the numerator and/or denominator, perform these operations before converting to a decimal equivalent.

To solve $\dfrac{1+5}{(6)(2)}$

First, perform the addition in the numerator and the multiplication in the denominator to get:

$$\frac{6}{12}$$

Then divide 6 by 12 to get 0.50.

Note that in statistics you should report decimal equivalents as answers and *not* fractions. Thus, you should report 0.50 and *not* 6/12 or 1/2.

If a number consists of a whole number and a fraction, it is known as a *mixed number*. For example:

5 1/7 is a mixed number that consists of 5 and 1/7.

Before working with mixed numbers, convert the fractional part to its decimal equivalent. For example, to convert 5 1/7 to its decimal equivalent, do this:

Divide 1 by 7 = 0.14.
Add it to the whole number: 5 + 0.14 = 5.14.
Thus, 5 1/7 equals 5.14.

Exercise for Section E

(Note: The answers to items 1–8 are given at the end of this supplement.)
1. "In 3/4, 3 is the denominator." TRUE or FALSE?
2. "In 3/4, 3 is the numerator." TRUE or FALSE?
3. "In 5/6, 6 is the denominator." TRUE or FALSE?

4. "Other things being equal, as the numerator increases, the value of the fraction decreases." TRUE or FALSE?

5. "Other things being equal, as the denominator increases, the value of the fraction decreases." TRUE or FALSE?

6. What is the decimal equivalent of 2/5?

7. What is the decimal equivalent of 1/12?

8. What is the decimal equivalent of 3/4?

9. "In 7/8, 7 is the numerator." TRUE or FALSE?

10. "In 7/8, 7 is the denominator." TRUE or FALSE?

11. "In 1/4, 4 is the numerator." TRUE or FALSE?

12. Other things being equal, as the denominator increases, what happens to the value of the fraction?

13. Other things being equal, as the numerator increases, what happens to the value of the fraction?

14. What is the decimal equivalent of 10/100?

15. What is the decimal equivalent of 1/3?

16. What is the decimal equivalent of 1/2?

Section F   Percentages

We often use percentages to describe data. A percentage represents a part of 100. For example:

25% stands for 25 out of 100.

50% stands for 50 out of 100.

By using percentages, we are using 100 as the base regardless of how many subjects we are describing. For example, if 10 out of 20 subjects are male, we would report that 50% are male. If, for another group, 15 out of 30 subjects are male, we would also report that 50% are male. By using a common base, percentages facilitate the comparison of groups of unequal size.

To convert a fraction to a percentage, divide the numerator by the denominator and multiply by 100 as in these examples:

$$1/3 = 0.333 \times 100 = 33.3\%$$
$$2/9 = 0.222 \times 100 = 22.2\%$$

In scientific writing, authors usually report percentages to one or two decimal places. Notice that in the answer to the following problem, the value in the tenths' place is zero:

$$3/4 = 0.750 \times 100 = 75.0\%$$

We show the zero in the tenths' place in order to be consistent in our reporting (e.g., if other percentages such as 33.3% have been reported) and to show that the answer is accurate to the tenths' place.

Note that when computing a percentage, you may simply move the decimal point two places to the right, which has the same effect as multiplying by 100. For example, for 1/3, divide 1 by 3 to get 0.333. Then move the decimal point two places to the right to get 33.3%.

Suppose that we are studying 79 voters, and 35 of them classified themselves as liberals. We could report that 35/79 are liberals. In scientific writing, however, authors usually would compute the corresponding percentage:

$$35 \text{ divided by } 79 = 0.443 \times 100 = 44.3\%$$

and report that:

$$44.3\% \ (N = 35) \text{ are liberals.}$$

N stands for *number of subjects or cases*. It is desirable when reporting percentages to also report the number of cases underlying each percentage. The number of cases is an important piece of information for your reader, as these examples illustrate:

Study A: 80% ($N = 5$) of the dentists recommend Brand X.

Study B: 80% ($N = 103$) of the dentists recommend Brand X.

Even though the percentages are the same in the two studies, the larger N in Study B indicates that the results of Study B are more reliable than the results of Study A.

The percentages for a group may not always sum to exactly 100%, as illustrated by the following example in Table 1. There were no computational errors. Instead, the total of 100.1% is slightly greater than 100% because of rounding. For example, for Method A, the precise percentage is 18.75%. Because this was rounded to 18.8%, a slight amount of error has been introduced. For most practical purposes, this error is of little consequence. Note that in the last column the upper-case P is used, which is a symbol for *percentage*.

Table 1
Number and Percentage of Teachers Who Prefer Each Method

Method	N	P
Method A	9	18.8%
Method B	18	37.5%
Method C	21	43.8%
TOTAL	48	100.1%

If you still do not feel comfortable computing percentages, consider the example of the teachers more carefully:

For Method A, first divide 9 by 48 = 0.1875.

Then multiply by 100, 0.1875 × 100 = 18.75%, which rounds to 18.8%.

For Method B, first divide 18 by 48 = 0.3750.

Then multiply by 100, 0.3750 × 100 = 37.50%, which rounds to 37.5%.

For Method C, first divide 21 by 48 = 0.4375.

Then multiply by 100, 0.4375 × 100 = 43.75, which rounds to 43.8%.

Exercise for Section F

(Note: The answers to items 1–8 are given at the end of this supplement.)

1. "25% stands for 25 out of 50." TRUE or FALSE?

2. "The base for percentages is 100." TRUE or FALSE?

3. What percentage corresponds to 1/5?

4. What percentage corresponds to 3/10?

5. If 95 subjects were studied and 71 of them expressed Opinion A, what percentage of them expressed Opinion A?

6. If 200 subjects were studied and 120 of them expressed a preference for Brand X, what percentage expressed a preference for Brand X?

7. "It is recommended that when you report a percentage, you should also report the corresponding value of N." TRUE or FALSE?

8. "In statistics, the lower-case p stands for *percentage*." TRUE or FALSE?

9. "11% stands for 11 out of 100." TRUE or FALSE?

10. "By using a common base of 100, percentages facilitate the comparison of groups of unequal size." TRUE or FALSE?

11. What percentage corresponds to 2/5?

12. What percentage corresponds to 1/100?

13. If 66 subjects were studied and 21 of them expressed Opinion X, what percentage of them expressed Opinion X?

14. If 462 subjects were studied and 99 of them expressed a preference for Brand Y, what percentage expressed a preference for Brand Y?

15. "When writing a scientific report, there is no need to report Ns if percentages are being reported." TRUE or FALSE?

16. "In statistics, the upper-case P stands for *percentage*." TRUE or FALSE?

Answers to Selected Questions

Section A: 1. 80, 2. 3, 3. 15, 4. 28, 5. 109, 6. 8, 7. 15, 8. 10, 9. 110, 10. 5, 11. A, 12. C.

Section B: 1. A, 2. A, 3. B, 4. C, 5. 12, 6. 10,000, 7. 15, 8. 324, 9. 54, 10. 36.

Section C: 1. –56, 2. –200, 3. –7, 4. –11, 5. 6, 6. 120, 7. –20, 8. –4, 9. –4, 10. –3, 11. 8, 12. –15, 13. –30, 14. 19, 15. 33, 16. –38, 17. B.

Section D: 1. F, 2. T, 3. F, 4. T, 5. T, 6. F, 7. F, 8. T, 9. T, 10. F, 11. T, 12. T.

Section E: 1. F, 2. T, 3. T, 4. F, 5. T, 6. 0.40, 7. 0.08, 8. 0.75.

Section F: 1. F, 2. T, 3. 20.0%, 4. 30.0%, 5. 74.7%, 6. 60.0%, 7. T, 8. F.

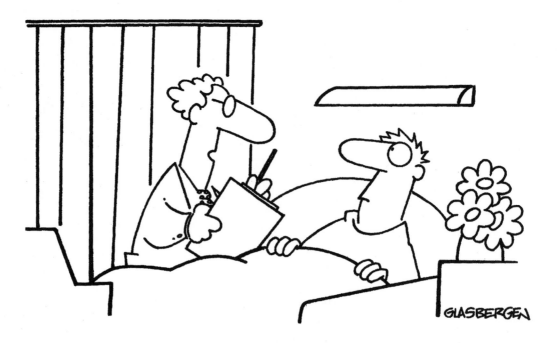

"I think you need a heart transplant, my associate thinks you need a bypass, and our statistician thinks you just need to be rebooted."

Appendix A
Computational Formulas for the
Standard Deviation

For the standard deviation of a population (i.e., all members of the population tested):

$$S = \frac{1}{N}\sqrt{N\Sigma X^2 - (\Sigma X)^2}$$

Where:
N = number of cases.
ΣX^2 = sum of the squared scores (sum after squaring each score).
$(\Sigma X)^2$ = square of the sum of the scores (sum and then square).

Example:

X	X^2
3	9
5	25
7	49
8	64
$\Sigma X = $ 23	$\Sigma X^2 = $ 147

$$S = \frac{1}{4}\sqrt{(4)(147) - (23)^2}$$
$$= .25\sqrt{588 - 529}$$
$$= .25\sqrt{59}$$
$$= (.25)(7.681) = 1.920 = 1.92$$

For the standard deviation of a population estimated from a sample (i.e., only a sample of the population tested), use the following formula, which is applied here to the scores shown above.[1]

$$s = \sqrt{\frac{N\Sigma X^2 - (\Sigma X)^2}{N(N-1)}} = \sqrt{\frac{(4)(147) - (23)^2}{4(4-1)}} = \sqrt{\frac{588 - 529}{12}} = \sqrt{\frac{59}{12}} = \sqrt{4.917} = 2.217 = 2.22$$

---

[1]The two formulas are applied to the same set of scores here for instructional purposes only. As you can see, the second formula gives a higher value for the standard deviation. This is because of an adjustment made by the second formula that takes into account the fact that extreme scores tend to be underestimated when small samples are used. With a large N, the two formulas yield almost identical results. Of course, in practice, only one formula should be applied to a given set of scores. If you are analyzing scores for an entire population, use the first formula; if you are analyzing those of a sample, use the second formula.

Appendix B
Notes on Interpreting the Pearson *r*

The value of a Pearson *r* can be misleadingly low for two reasons. First, its value is diminished if the variability in a group is artificially low. For the sake of illustration, let's assume that we wanted to study the relationship between height and weight in the adult population but, foolishly, selected only subjects who were exactly six feet tall. When weighing them, we found some variation in their weights. What is the correlation between height and weight among such a group? Even though there is a positive relationship between the two variables in the general adult population, the correlation in this odd sample is zero. This must be the result because those who weigh more and those who weigh less are all of the same height. (This sample cannot show that those who are taller tend to weigh more because all subjects are of the same height. Thus, the value of the Pearson *r* will equal 0.00.) A more realistic example is the relationship between scores on a college admissions test and grades earned in college. Although the test is given to all applicants in order to make admissions decisions about all of them, grades are available only for those who were admitted, and the correlation between scores and grades can be computed only for those subjects on whom we have complete data. Those for whom we have complete data are those with higher but less variable scores. Thus, the value of the Pearson *r* will be lower than would be obtained if we correlated using scores and grades for *all* applicants.

Second, an *r* can be misleadingly low if the underlying relationship is curvilinear. For example, the relationship between test-taking anxiety and performance on standardized tests might be curvilinear. That is, small amounts of anxiety might be beneficial in motivating subjects to do well on a test, but larger amounts might be detrimental. Thus, as anxiety increases, up to a point, there is a positive relationship with anxiety; after reaching a critical point, as anxiety increases, there is a negative relationship. If the Pearson *r* is computed for such data, the negative part of the relationship will cancel out the positive part, yielding an *r* of near zero. Pearson recognized this problem and warned against using his statistic for describing curvilinear relationships. Other techniques such as the *correlation ratio*, which are beyond the scope of this book, are available for describing curvilinear relationships. Fortunately, such relationships are relatively rare in the social and behavioral sciences.

Appendix C
Definition Formula for the Pearson *r*

After studying Sections 21 and 22, you may wonder how it is possible to correlate variables that have such diverse scales as GPAs (0.00 to 4.00) and *SAT* scores (200 to 800) and, regardless of the scale, always obtain a value of *r* that always ranges from −1.00 to 1.00. At first, it may seem like we are comparing apples and oranges. However, the answer is rather straightforward if you recall the characteristics of standard scores (i.e., *z*-scores).[1] For example, a person who has a GPA that is average for the group has a *z*-score of 0.00 on GPA; if the same person has an *SAT* that is also average for the group, her *z*-score is also 0.00 for *SAT*. In other words, *z*-scores make it possible to compare scores that were *not* directly comparable in their original form. Taking advantage of this property, Pearson defined a perfect, direct correlation as one in which each person had the same *z*-score on both variables. A perfect, inverse correlation is defined as one in which each person has the exact opposite *z*-score on each variable (e.g., a person with a *z*-score of 1.50 on one variable has a −1.50 on the other). This definition formula for the Pearson *r* expressed in terms of *z*-scores is:

$$r = \frac{\Sigma \left(z_x z_y \right)}{N}$$

Thus, in order to use this formula, first compute the *z*-score for each subject on variable *X* and then on variable *Y*. Multiply the corresponding *z*-scores and sum their products. Finally, divide by *N*.

The computational formula for *r* presented in Section 22 is algebraically equivalent to the one shown above. However, the one in Section 22 eliminates the need to calculate the *z*-scores for each subject, which would be a major task if there were a large number of subjects.

---

[1]See Section 16 for a thorough review.

Appendix D
Spearman's *rho*

A statistician named Spearman developed a formula for calculating a product-moment correlation coefficient that is easier to compute than the Pearson r when subjects are ranked on both variables. Usually, a rank of 1 is given to the subject with the best performance, a rank of 2 to the subject with the next best performance, etc.

When using the following formula, the result is usually referred to as *rho* or r_s.

$$rho = 1 - \frac{6 \Sigma D^2}{N(N^2 - 1)}$$

Where:

      D is the difference between the two ranks for each subject.

      N is the number of subjects.

      Both 1 and 6 are constants.

Application of the formula is illustrated for the ranks in columns 2 and 3 below.

Col. 1	Col. 2	Col. 3	Col. 4	Col. 5
			Difference	Squared Difference
Subject	Rank on X	Rank on Y	D	D^2
Alexis	1	3	2	4
Allen	5	4	1	1
Bob	2	1	1	1
Mary	3	5	2	4
Joe	4	2	2	4
				$\Sigma D^2 = 14$

$$rho = 1 - \frac{(6)(14)}{(5)(5^2 - 1)} = 1 - \frac{84}{(5)(25 - 1)} = 1 - \frac{84}{120} = 1 - .70 = .30$$

Application of the formula becomes more difficult if there are ties (e.g., two subjects tied as number 1 on the same variable). Dealing with ties is not covered here. However, you may obtain the correct answer when there are ties by applying the formula for the Pearson r, which is illustrated in Section 22. In other words, simply treat the ranks as though they were scores and apply the formula for the Pearson *r*.

Spearman's *rho* was developed as a shortcut computational procedure for a specialized situation (when there are ranks) before there were handheld calculators or computers. With the advent of modern computing devices, *rho* is not as important as it once was because the computation of the Pearson *r* is fairly easy with calculators and computers.

Because *rho* is a product-moment correlation, for all practical purposes it is interpreted in the same way as a Pearson *r*.

Appendix E
Standard Error of Estimate

When the correlation between a predictor variable (called X) and the variable being predicted (called Y or the criterion variable) is less than perfect, as it usually is, we will make errors when we predict from X to Y. Thus, we should allow for a margin of error when making predictions. To do this, first compute the ***standard error of estimate*** (S_{yx}) using the following formula. Its application is shown using the data from Table 26.1 for which there is a correlation of $-.77$ between X and Y.

$$S_{yx} = \sqrt{\frac{\left[\Sigma Y^2 - \dfrac{(\Sigma Y)^2}{N}\right] - \left[\Sigma XY - \dfrac{(\Sigma X)(\Sigma Y)}{N}\right]^2 \div \left[\Sigma X^2 - \dfrac{(\Sigma X)^2}{N}\right]}{N-2}}$$

$$= \sqrt{\frac{\left[207 - \dfrac{31^2}{6}\right] - \left[67 - \dfrac{(19)(31)}{6}\right]^2 \div \left[95 - \dfrac{19^2}{6}\right]}{6-2}}$$

$$= \sqrt{\frac{46.833 - (971.382 \div 34.833)}{4}} = \sqrt{\frac{18.946}{4}} = \sqrt{4.7365} = 2.176 = 2.18$$

The result, 2.18, is an estimate of the standard deviation of the scatter around the prediction line. Within one standard deviation, we expect to find about 68% of the cases in a normal distribution. Thus, for the data we are considering, we can say that we have 68% confidence that the true score a person will earn will be within 2.18 points of the predicted score. For example, in Section 26 we saw that for the above data, a person with a score of 6 on X has a predicted score on Y of 2.66. To apply the rule for using the standard error of estimate, do the following: Subtract the standard error of estimate from the predicted score (2.66 − 2.18 = 0.48) and add it to the predicted score (2.66 + 2.18 = 4.84). The result is a confidence interval. We can now say that we have 68% confidence that the true score a person will earn will be in the range of 0.48 and 4.84.

It is beyond the scope of this appendix to explore the theory underlying the standard error of estimate. However, the theoretical underpinnings are analogous to those underlying the standard error of the mean, which is explored in detail in Section 31.

Appendix F
Standard Error of a Median and a Percentage

The **standard error of a median** is greater than the standard error of the mean as indicated by this formula:

$$se \ (median) = \frac{1.253s}{\sqrt{n}}$$

The formula is easy to use. Just multiply the standard deviation (s) by 1.253 and divide by the square root of n.

The **standard error of a percentage** is defined by this formula:

$$se \ (percentage) = \sqrt{\frac{PQ}{N}}$$

Where:

P is the percentage in question.

$Q = 100 - P$.

N is the number of cases.

Appendix G
Confidence Interval for the Mean: Small Samples

When using small sample sizes, such as 60 or less, the sampling distribution of the mean is not normal and the constants of 1.96 (for 95% confidence) and 2.58 (for 99% confidence) will lead to inaccurate results. A statistician named W. S. Gossett developed the t table in Table 3, which can be used to obtain the appropriate constants. To use the table, first compute a statistic known as degrees of freedom (df), using this formula:

$$df = n - 1$$

Thus, if $n = 11$, then $df = 10$. Look up the df of 10 in the first column of Table 3. To get the constant for the 95% confidence interval, go to the right to the second column labeled *95% C.I.* There you will find a constant of 2.228. (Note that this is larger than the constant of 1.96 that you learned about in Section 32.)

Multiply the standard error of the mean by the constant that you find in Table 3. Thus, if $SE_M = 2.00$, then $2.00 \times 2.228 = 4.456$. Add 4.456 to the mean and subtract 4.456 from the mean in order to obtain the 95% confidence interval. Thus, if $M = 100.00$, the limits of the 95% confidence interval equal 95.544 and 104.456.

For a 99% confidence interval with a df of 10, use the constant 3.169 found in the third column of Table 3, labeled *99% C.I.*

"Your résumé is bloated with half-truths, false praise, exaggeration, and unsubstantiated accomplishments. I'd like to hire you to prepare the statistical charts for our Annual Report."

Appendix H
Computation of the Precise Median

In Section 11, you learned how to compute the median. When there are ties in the middle of the distribution, the precise median can be obtained by first putting the scores in a frequency distribution and then applying the formula shown below.[1]

Score	Frequency (f)	Cumulative Frequency (cf)
20	2	22
19	0	20
18	4	20
17	6	16
16	5	10
15	3	5
14	2	2
	$N = 22$	

The formula is:

$$\text{Median} = L + \frac{N(.5) - cfb}{f}$$

Where:

L is the exact lower limit of the interval that contains the median; this is the *interval of interest*. To find this interval, divide N by 2 (in this case, $22/2 = 11$), and go *up* the cf column looking for subject number 11. The cf column indicates that up to a score of 16, there are 10 subjects. The next score interval (17) contains subjects 11 through 16. Since we are looking for subject 11, **17** is the score interval of interest. Subtract .5 from the score of interest to obtain the exact lower limit (i.e., $17 - .5 = \mathbf{16.5}$).

---

[1] Notice that for all scores, except 19, there are ties, as indicated by frequencies (f) of more than one. Section 11 shows how to compute the precise median when there are no ties near the median. The following method will yield the precise median whether or not there are ties.

N is the total number of cases; in this example, it is **22**.

cfb is the cumulative frequency (*cf*) for the score immediately *below* the *interval of interest* (in this case, *cfb* = **10** for the score of 16, which is the score below 16.5).

f is the frequency of the interval of interest (in this case, **6** for the interval associated with a score of 17).

Substituting into the formula, we obtain the following:

$$\text{Median} = 16.5 + \frac{(22)(.5) - 10}{6} = 16.5 + \frac{1}{6} = 16.5 + .167 = 16.667$$

BOOKS

"Yes, we have *Chicken Soup for the Statistics Professor's Soul.*
The price is $(4)(SD) + M - r^2/t + (F)(df)(12)/100 - z + T - p$."

Table 1
Abbreviated Table of the Normal Curve

z-score	Col. 2 % of cases from M to z	Col. 3 % of cases larger part	Col 4. % of cases smaller part	z-score	Col. 2 % of cases from M to z	Col. 3 % of cases larger part	Col. 4 % of cases smaller part	z-score	Col. 2 % of cases from M to z	Col. 3 % of cases larger part	Col. 4 % of cases smaller part
0.00	0.00	50.00	50.00	1.20	38.49	88.49	11.51	2.40	49.18	99.18	0.82
0.05	1.99	51.99	48.01	1.25	39.44	89.44	10.56	2.45	49.29	99.29	0.71
0.10	3.98	53.98	46.02	1.30	40.32	90.32	9.68	2.50	49.38	99.38	0.62
0.15	5.96	55.96	44.04	1.35	41.15	91.15	8.85	2.55	49.46	99.46	0.54
0.20	7.93	57.93	42.07	1.40	41.92	91.92	8.08	2.60	49.53	99.53	0.47
0.25	9.87	59.87	40.13	1.45	42.65	92.65	7.35	2.65	49.60	99.60	0.40
0.30	11.79	61.79	38.21	1.50	43.32	93.32	6.68	2.70	49.65	99.65	0.35
0.35	13.68	63.68	36.32	1.55	43.94	93.94	6.06	2.75	49.70	99.70	0.30
0.40	15.54	65.54	34.46	1.60	44.52	94.52	5.48	2.80	49.74	99.74	0.26
0.45	17.36	67.36	32.64	1.65	45.05	95.05	4.95	2.85	49.78	99.78	0.22
0.50	19.15	69.15	30.85	1.70	45.54	95.54	4.46	2.90	49.81	99.81	0.19
0.55	20.88	70.88	29.12	1.75	45.99	95.99	4.01	2.95	49.84	99.84	0.16
0.60	22.57	72.57	27.43	1.80	46.41	96.41	3.59	3.00	49.87	99.87	0.13
0.65	24.22	74.22	25.78	1.85	46.48	96.78	3.22	**Values of Special Interest**			
0.70	25.80	75.80	24.20	1.90	47.13	97.13	2.87	1.96	47.50	97.50	2.50
0.75	27.34	77.34	22.66	1.95	47.44	97.44	2.56	2.58	49.51	99.51	0.49
0.80	28.81	78.81	21.19	2.00	47.72	97.72	2.28				
0.85	30.23	80.23	19.77	2.05	47.98	97.98	2.02				
0.90	31.59	81.59	18.41	2.10	48.21	98.21	1.79				
0.95	32.89	82.89	17.11	2.15	48.42	98.42	1.58				
1.00	34.13	84.13	15.87	2.20	48.61	98.61	1.39				
1.05	35.31	85.31	14.69	2.25	48.78	98.78	1.22				
1.10	36.43	86.43	13.57	2.30	48.93	98.93	1.07				
1.15	37.49	87.49	12.51	2.35	49.06	99.06	0.94				

Table 2
Table of Random Numbers

Row #																		
1	2	1	0	4	9	8	0	8	8	8	0	6	9	2	4	8	2	6
2	0	7	3	0	2	9	4	8	2	7	8	9	8	9	2	9	7	1
3	4	4	9	0	0	2	8	6	2	6	7	7	7	3	1	2	5	1
4	7	3	2	1	1	2	0	7	7	6	0	3	8	3	4	7	8	1
5	3	3	2	5	8	3	1	7	0	1	4	0	7	8	9	3	7	7
6	6	1	2	0	5	7	2	4	4	0	0	6	3	0	2	8	0	7
7	7	0	9	3	3	3	7	4	0	4	8	8	9	3	5	8	0	5
8	7	5	1	9	0	9	1	5	2	6	5	0	9	0	3	5	8	8
9	3	5	6	9	6	5	0	1	9	4	6	6	7	5	6	8	3	1
10	8	5	0	3	9	4	3	4	0	6	5	1	7	4	4	6	2	7
11	0	5	9	6	8	7	4	8	1	5	5	0	5	1	7	1	5	8
12	7	6	2	2	6	9	6	1	9	7	1	1	4	7	1	6	2	0
13	3	8	4	7	8	9	8	2	2	1	6	3	8	7	0	4	6	1
14	1	9	1	8	4	5	6	1	8	1	2	4	4	4	2	7	3	4
15	1	5	3	6	7	6	1	8	4	3	1	8	8	7	7	6	0	4
16	0	5	5	3	6	0	7	1	3	8	1	4	6	7	0	4	3	5
17	2	2	3	8	6	0	9	1	9	0	4	4	7	6	8	1	5	1
18	2	3	3	2	5	5	7	6	9	4	9	7	1	3	7	9	3	8
19	8	5	5	0	5	3	7	8	5	4	5	1	6	0	4	8	9	1
20	0	6	1	1	3	4	8	6	4	3	2	9	4	3	8	7	4	1
21	9	1	1	8	2	9	0	6	9	6	9	4	2	9	9	0	6	0
22	3	7	8	0	6	3	7	1	2	6	5	2	7	6	5	6	5	1
23	5	3	0	5	1	2	1	0	9	1	3	7	5	6	1	2	5	0
24	7	2	4	8	6	7	9	3	8	7	6	0	9	1	6	5	7	8
25	0	9	1	6	7	0	3	8	0	9	1	5	4	2	3	2	4	5
26	3	8	1	4	3	7	9	2	4	5	1	2	8	7	7	4	1	3

Table 3
Constants Based on *t* for Computing Confidence Intervals for the Mean Based on Small Samples[1]

df	95% C.I.	99% C.I.	df	95% C.I.	99% C.I.
1	12.706	63.657	24	2.064	2.797
2	4.303	9.925	25	2.060	2.787
3	3.182	5.841	26	2.056	2.779
4	2.776	4.604	27	2.052	2.771
5	2.571	4.032	28	2.048	2.763
6	2.447	3.707	29	2.045	2.756
7	2.365	3.499	30	2.042	2.750
8	2.306	3.355	40	2.021	2.704
9	2.262	3.250	60	2.000	2.660
10	2.228	3.169	120	1.980	2.617
11	2.201	3.106	Infinity	1.960	2.576
12	2.179	3.055			
13	2.160	3.012			
14	2.145	2.977			
15	2.131	2.947			
16	2.120	2.921			
17	2.110	2.898			
18	2.101	2.878			
19	2.093	2.861			
20	2.086	2.845			
21	2.080	2.831			
22	2.074	2.819			
23	2.069	2.807			

[1]Abridged from Fisher, R. A., & Yates, F. *Statistical tables for biological, agricultural, and medical research.* Edinburgh: Oliver Boyd Ltd.

Table 4
Critical Values of *t* for Two-Tailed *t* Test[1]

df	.05 level	.01 level	.001 level	*df*	.05 level	.01 level	.001 level
1	12.706	63.657	636.619	24	2.064	2.797	3.745
2	4.303	9.925	31.598	25	2.060	2.787	3.725
3	3.182	5.841	12.941	26	2.056	2.779	3.707
4	2.776	4.604	8.610	27	2.052	2.771	3.690
5	2.571	4.032	6.859	28	2.048	2.763	3.674
6	2.447	3.707	5.959	29	2.045	2.756	3.659
7	2.365	3.499	5.405	30	2.042	2.750	3.646
8	2.306	3.355	5.041	40	2.021	2.704	3.551
9	2.262	3.250	4.781	60	2.000	2.660	3.460
10	2.228	3.169	4.587	120	1.980	2.617	3.373
11	2.201	3.106	4.437	Infinity	1.960	2.576	3.291
12	2.179	3.055	4.318				
13	2.160	3.012	4.221				
14	2.145	2.977	4.140				
15	2.131	2.947	4.073				
16	2.120	2.921	4.015				
17	2.110	2.898	3.965				
18	2.101	2.878	3.922				
19	2.093	2.861	3.883				
20	2.086	2.845	3.850				
21	2.080	2.831	3.819				
22	2.074	2.819	3.792				
23	2.069	2.807	3.767				

[1]Abridged from Fisher, R. A., & Yates, F. *Statistical tables for biological, agricultural, and medical research.* Edinburgh: Oliver Boyd Ltd.

Table 5
Critical Values of t for One-Tailed t Test[1]

df	.05 level	.01 level	.001 level	df	.05 level	.01 level	.001 level
1	6.314	31.821	318.310	24	1.711	2.492	3.467
2	2.920	6.965	22.326	25	1.708	2.485	3.450
3	2.353	4.541	10.213	26	1.706	2.479	3.435
4	2.132	3.747	7.173	27	1.703	2.473	3.421
5	2.015	3.365	5.893	28	1.701	2.467	3.408
6	1.943	3.143	5.208	29	1.699	2.462	3.396
7	1.895	2.998	4.785	30	1.697	2.457	3.385
8	1.860	2.896	4.501	40	1.684	2.423	3.307
9	1.833	2.821	4.297	60	1.671	2.390	3.232
10	1.812	2.764	4.144	120	1.658	2.358	3.160
11	1.796	2.718	4.025	Infinity	1.645	2.326	3.090
12	1.782	2.681	3.930				
13	1.771	2.650	3.852				
14	1.761	2.624	3.787				
15	1.753	2.602	3.733				
16	1.746	2.583	3.686				
17	1.740	2.567	3.646				
18	1.734	2.552	3.610				
19	1.729	2.539	3.579				
20	1.725	2.528	3.552				
21	1.721	2.518	3.527				
22	1.717	2.508	3.505				
23	1.714	2.500	3.485				

[1]Abridged from Fisher, R. A., & Yates, F. *Statistical tables for biological, agricultural, and medical research.* Edinburgh: Oliver Boyd Ltd.

Table 6
Critical Values of *F* for the .05 Level

Within Groups Degrees of Freedom	Between Groups Degrees of Freedom (Numerator)												
	1	2	3	4	5	6	7	8	9	10	11	12	14
1	161	200	216	225	230	234	237	239	241	242	243	244	245
2	18.51	19.00	19.16	19.25	19.30	19.33	19.36	19.37	19.38	19.39	19.40	19.41	19.42
3	10.13	9.55	9.28	9.12	9.01	8.94	8.88	8.84	8.81	8.78	8.76	8.74	8.71
4	7.71	6.94	6.59	6.39	6.26	6.16	6.09	6.04	6.00	5.96	5.93	5.91	5.87
5	6.61	5.79	5.41	5.19	5.05	4.95	4.88	4.82	4.78	4.74	4.70	4.68	4.64
6	5.99	5.14	4.76	4.53	4.39	4.28	4.21	4.15	4.10	4.06	4.03	4.00	3.96
7	5.59	4.74	4.35	4.12	3.97	3.87	3.79	3.73	3.68	3.63	3.60	3.57	3.52
8	5.32	4.46	4.07	3.84	3.69	3.58	3.50	3.44	3.39	3.34	3.31	3.28	3.23
9	5.12	4.26	3.86	3.63	3.48	3.37	3.29	3.23	3.18	3.13	3.10	3.07	3.02
10	4.96	4.10	3.71	3.48	3.33	3.22	3.14	3.07	3.02	2.97	2.94	2.91	2.86
11	4.84	3.98	3.59	3.36	3.20	3.09	3.01	2.95	2.90	2.86	2.82	2.79	2.74
12	4.75	3.88	3.49	3.26	3.11	3.00	2.92	2.85	2.80	2.76	2.72	2.69	2.64
13	4.67	3.80	3.41	3.18	3.02	2.92	2.84	2.72	2.77	2.63	2.63	2.60	2.55
14	4.60	3.74	3.34	3.11	2.96	2.85	2.77	2.70	2.65	2.60	2.56	2.53	2.48
15	4.54	3.68	3.29	3.06	2.90	2.79	2.70	2.64	2.59	2.55	2.51	2.48	2.43
16	4.49	3.63	3.24	3.01	2.85	2.74	2.66	2.59	2.54	2.49	2.45	2.42	2.37
17	4.45	3.59	3.20	2.96	2.81	2.70	2.62	2.55	2.50	2.45	2.41	2.38	2.33
18	4.41	3.55	3.16	2.93	2.77	2.66	2.58	2.51	2.46	2.41	2.37	2.34	2.29
19	4.38	3.52	3.13	2.90	2.74	2.63	2.55	2.48	2.43	2.38	2.34	2.31	2.26
20	4.35	3.49	3.10	2.87	2.71	2.60	2.52	2.45	2.40	2.35	2.31	2.28	2.23
21	4.32	3.47	3.07	2.84	2.68	2.57	2.49	2.42	2.37	2.32	2.28	2.25	2.20
22	4.30	3.44	3.05	2.82	2.66	2.55	2.47	2.40	2.35	2.30	2.26	2.23	2.18
23	4.28	3.42	3.03	2.80	2.64	2.53	2.45	2.38	2.32	2.28	2.24	2.20	2.14
24	4.26	3.40	3.01	2.78	2.62	2.51	2.43	2.36	2.30	2.26	2.22	2.18	2.13
25	4.24	3.38	2.99	2.76	2.60	2.49	2.41	2.34	2.28	2.24	2.20	2.16	2.11
26	4.22	3.37	2.98	2.74	2.59	2.47	2.39	2.32	2.27	2.22	2.18	2.15	2.10
27	4.21	3.35	2.96	2.73	2.57	2.46	2.37	2.30	2.25	2.20	2.16	2.13	2.08
28	4.20	3.34	2.95	2.71	2.56	2.44	2.36	2.29	2.24	2.19	2.15	2.12	2.06
29	4.18	3.33	2.93	2.70	2.54	2.43	2.35	2.28	2.22	2.18	2.14	2.10	2.05

Table 6 (Continued)

Between Groups Degrees of Freedom (Numerator)											Within Groups Degrees of Freedom
16	20	24	30	40	50	75	100	200	500	infinity	
246	248	249	250	251	252	253	253	254	254	254	1
19.43	19.44	19.45	19.46	19.47	19.47	19.48	19.49	19.49	19.50	19.50	2
8.69	8.66	8.64	8.62	8.60	8.58	8.57	8.56	8.54	8.54	8.53	3
5.84	5.80	5.77	5.74	5.71	5.70	5.68	5.66	5.65	5.64	5.63	4
4.60	4.56	4.53	4.50	4.46	4.44	4.42	4.40	4.38	4.37	4.36	5
3.92	3.87	3.84	3.81	3.77	3.75	3.72	3.71	3.69	3.68	3.67	6
3.49	3.44	3.41	3.38	3.34	3.32	3.29	3.28	3.25	3.24	3.23	7
3.20	3.15	3.12	3.08	3.05	3.03	3.00	2.98	2.96	2.94	2.93	8
2.98	2.93	2.90	2.86	2.82	2.80	2.77	2.76	2.73	2.72	2.71	9
2.82	2.77	2.74	2.70	2.67	2.64	2.61	2.59	2.56	2.55	2.54	10
2.70	2.65	2.61	2.57	2.53	2.50	2.47	2.45	2.42	2.41	2.40	11
2.60	2.54	2.50	2.46	2.42	2.40	2.36	2.35	2.32	2.31	2.30	12
2.51	2.46	2.42	2.38	2.34	2.32	2.28	2.26	2.24	2.22	2.21	13
2.44	2.39	2.35	2.31	2.27	2.24	2.21	2.19	2.16	2.14	2.13	14
2.39	2.33	2.29	2.25	2.21	2.18	2.15	2.12	2.10	2.08	2.07	15
2.33	2.28	2.24	2.20	2.16	2.13	2.09	2.07	2.04	2.02	2.01	16
2.29	2.23	2.19	2.15	2.11	2.08	2.04	2.02	1.99	1.97	1.96	17
2.25	2.19	2.15	2.11	2.07	2.04	2.00	1.98	1.95	1.93	1.92	18
2.21	2.15	2.11	2.07	2.02	2.00	1.96	1.94	1.91	1.90	1.88	19
2.18	2.12	2.08	2.04	1.99	1.96	1.92	1.90	1.87	1.85	1.84	20
2.15	2.09	2.05	2.00	1.96	1.93	1.89	1.87	1.84	1.82	1.81	21
2.13	2.07	2.03	1.98	1.93	1.91	1.87	1.84	1.81	1.80	1.78	22
2.10	2.04	2.00	1.96	1.91	1.88	1.84	1.82	1.79	1.77	1.76	23
2.09	2.02	1.98	1.94	1.89	1.86	1.82	1.80	1.76	1.74	1.73	24
2.06	2.00	1.96	1.92	1.87	1.84	1.80	1.77	1.74	1.72	1.71	25
2.05	1.99	1.95	1.90	1.85	1.82	1.78	1.76	1.72	1.70	1.69	26
2.03	1.97	1.93	1.88	1.84	1.80	1.76	1.74	1.71	1.68	1.67	27
2.02	1.96	1.91	1.87	1.81	1.78	1.75	1.72	1.69	1.67	1.65	28
2.00	1.94	1.90	1.85	1.80	1.77	1.73	1.71	1.68	1.65	1.64	29

Continued

Table 6 (Continued)

Within Groups Degrees of Freedom	Between Groups Degrees of Freedom (Numerator)												
	1	2	3	4	5	6	7	8	9	10	11	12	14
30	4.17	3.32	2.92	2.69	2.53	2.42	2.34	2.27	2.21	2.16	2.12	2.09	2.04
32	4.15	3.30	2.90	2.67	2.51	2.40	2.32	2.25	2.19	2.14	2.10	2.07	2.02
34	4.13	3.28	2.88	2.65	2.49	2.38	2.30	2.23	2.17	2.12	2.08	2.05	2.00
36	4.11	3.26	2.86	2.63	2.48	2.36	2.28	2.21	2.15	2.10	2.06	2.03	1.98
38	4.10	3.25	2.85	2.62	2.46	2.35	2.26	2.19	2.14	2.09	2.05	2.02	1.96
40	4.08	3.23	2.84	2.61	2.45	2.34	2.25	2.18	2.12	2.07	2.04	2.00	1.95
42	4.07	3.22	2.83	2.59	2.44	2.32	2.24	2.17	2.11	2.06	2.02	1.99	1.94
44	4.06	3.21	2.82	2.58	2.43	2.31	2.23	2.16	2.10	2.05	2.01	1.98	1.92
46	4.05	3.20	2.81	2.57	2.42	2.30	2.22	2.14	2.09	2.04	2.00	1.97	1.91
48	4.04	3.19	2.80	2.56	2.41	2.30	2.21	2.14	2.08	2.03	1.99	1.96	1.90
50	4.03	3.18	2.79	2.56	2.40	2.29	2.20	2.13	2.07	2.02	1.98	1.95	1.90
55	4.02	3.17	2.78	2.54	2.38	2.27	2.18	2.11	2.05	2.00	1.97	1.93	1.88
60	4.00	3.15	2.76	2.52	2.37	2.25	2.17	2.10	2.04	1.99	1.95	1.92	1.86
65	3.99	3.14	2.75	2.51	2.36	2.24	2.15	2.08	2.02	1.98	1.94	1.90	1.85
70	3.98	3.13	2.74	2.50	2.35	2.23	2.14	2.07	2.01	1.97	1.93	1.89	1.84
80	3.96	3.11	2.72	2.48	2.33	2.21	2.12	2.05	1.99	1.95	1.91	1.88	1.82
100	3.94	3.09	2.70	2.46	2.30	2.19	2.10	2.03	1.97	1.92	1.88	1.85	1.79
125	3.92	3.07	2.68	2.44	2.29	2.17	2.08	2.01	1.95	1.90	1.86	1.83	1.77
150	3.91	3.06	2.67	2.43	2.27	2.16	2.07	2.00	1.94	1.89	1.85	1.82	1.76
200	3.89	3.04	2.65	2.41	2.26	2.14	2.05	1.98	1.92	1.87	1.83	1.80	1.74
400	3.86	3.02	2.62	2.39	2.23	2.12	2.03	1.96	1.90	1.85	1.81	1.78	1.72
1000	3.85	3.00	2.61	2.38	2.22	2.10	2.02	1.95	1.89	1.84	1.80	1.76	1.70
Infinity	3.84	2.99	2.60	2.37	2.21	2.09	2.01	1.94	1.88	1.83	1.79	1.75	1.69

Table 6 (Continued)

Between Groups Degrees of Freedom (Numerator)											Within Groups Degrees of Freedom
16	20	24	30	40	50	75	100	200	500	Infinity	
1.99	1.93	1.89	1.84	1.79	1.76	1.72	1.69	1.66	1.64	1.62	30
1.97	1.91	1.86	1.82	1.76	1.74	1.69	1.67	1.64	1.61	1.59	32
1.95	1.89	1.84	1.80	1.74	1.71	1.67	1.64	1.61	1.59	1.57	34
1.93	1.87	1.82	1.78	1.72	1.69	1.65	1.62	1.59	1.56	1.55	36
1.92	1.85	1.80	1.76	1.71	1.67	1.63	1.60	1.57	1.54	1.53	38
1.90	1.84	1.79	1.74	1.69	1.66	1.61	1.59	1.55	1.53	1.51	40
1.89	1.82	1.78	1.73	1.68	1.64	1.60	1.57	1.54	1.51	1.49	42
1.88	1.81	1.76	1.72	1.66	1.63	1.58	1.56	1.52	1.50	1.48	44
1.87	1.80	1.75	1.71	1.65	1.62	1.57	1.54	1.51	1.48	1.46	46
1.86	1.79	1.74	1.70	1.64	1.61	1.56	1.53	1.50	1.47	1.45	48
1.85	1.78	1.74	1.69	1.63	1.60	1.55	1.52	1.48	1.46	1.44	50
1.83	1.76	1.72	1.67	1.61	1.58	1.52	1.50	1.46	1.43	1.41	55
1.81	1.75	1.70	1.65	1.59	1.56	1.50	1.48	1.44	1.41	1.39	60
1.80	1.73	1.68	1.63	1.57	1.54	1.49	1.46	1.42	1.39	1.37	65
1.79	1.72	1.67	1.62	1.56	1.53	1.47	1.45	1.40	1.37	1.35	70
1.77	1.70	1.65	1.60	1.54	1.51	1.45	1.42	1.38	1.35	1.32	80
1.75	1.68	1.63	1.57	1.51	1.48	1.42	1.39	1.34	1.30	1.28	100
1.72	1.65	1.60	1.55	1.49	1.45	1.39	1.36	1.31	1.27	1.25	125
1.71	1.64	1.59	1.54	1.47	1.44	1.37	1.34	1.29	1.25	1.22	150
1.69	1.62	1.57	1.52	1.45	1.42	1.35	1.32	1.26	1.22	1.19	200
1.67	1.60	1.54	1.49	1.42	1.38	1.32	1.28	1.22	1.16	1.13	400
1.65	1.58	1.53	1.47	1.41	1.36	1.30	1.26	1.19	1.13	1.08	1000
1.64	1.57	1.52	1.46	1.40	1.35	1.28	1.24	1.17	1.11	1.00	Infinity

Table 7
Critical Values of *F* for the .01 Level

Within Groups Degrees of Freedom	Between Groups Degrees of Freedom (Numerator)												
	1	2	3	4	5	6	7	8	9	10	11	12	14
1	4,052	4,999	5,403	5,625	5,764	5,859	5,928	5,981	6,022	6,056	6,082	6,106	6,142
2	98.49	99.00	99.17	99.25	99.30	99.33	99.34	99.36	99.38	99.40	99.41	99.42	99.43
3	34.12	30.82	29.46	28.71	28.24	27.91	29.67	27.49	27.34	27.23	27.13	27.05	26.92
4	21.20	18.00	16.69	15.98	15.52	15.21	14.98	14.80	14.66	14.54	14.45	14.37	14.24
5	16.26	13.27	12.06	11.39	10.97	10.67	10.45	10.27	10.15	10.05	9.96	9.89	9.77
6	13.74	10.92	9.78	9.15	8.75	8.47	8.26	8.10	7.98	7.87	7.79	7.72	7.60
7	12.25	9.55	8.45	7.85	7.46	7.19	7.00	6.84	6.71	6.62	6.54	6.47	6.35
8	11.26	8.65	7.59	7.01	6.63	6.37	6.19	6.03	5.91	5.82	5.74	5.67	5.56
9	10.56	8.02	6.99	6.42	6.06	5.80	5.62	5.47	5.35	5.26	5.18	5.11	5.00
10	10.04	7.56	6.55	5.99	5.64	5.39	5.21	5.06	4.95	4.85	4.78	4.71	4.60
11	9.65	7.20	6.22	5.67	5.32	5.07	4.88	4.74	4.63	4.54	4.46	4.40	4.29
12	9.33	6.93	5.95	5.41	5.06	4.82	4.65	4.50	4.39	4.30	4.22	4.16	4.05
13	9.07	6.70	5.74	5.20	4.86	4.62	4.44	4.30	4.19	4.10	4.02	3.96	3.85
14	8.86	6.51	5.56	5.03	4.69	4.46	4.28	4.14	4.03	3.94	3.86	3.80	3.70
15	8.68	6.36	5.42	4.89	4.56	4.32	4.14	4.00	3.89	3.80	3.73	3.67	3.56
16	8.53	6.23	5.29	4.77	4.44	4.20	4.03	3.89	3.78	3.69	3.61	3.55	3.45
17	8.40	6.11	5.18	4.67	4.34	4.10	3.93	3.79	3.68	3.59	3.52	3.45	3.35
18	8.28	6.01	5.09	4.58	4.25	4.01	3.85	3.71	3.60	3.51	3.44	3.37	3.27
19	8.18	5.93	5.01	4.50	4.17	3.94	3.77	3.63	3.52	3.43	3.36	3.30	3.19
20	8.10	5.85	4.94	4.43	4.10	3.87	3.71	3.56	3.45	3.37	3.30	3.23	3.13
21	8.02	5.78	4.87	4.37	4.04	3.81	3.65	3.51	3.40	3.31	3.24	3.17	3.07
22	7.94	5.72	4.82	4.31	3.99	3.76	3.59	3.45	3.35	3.26	3.18	3.12	3.02
23	7.88	5.66	4.76	4.26	3.94	3.71	3.54	3.41	3.30	3.21	3.14	3.07	2.97
24	7.82	5.61	4.72	4.22	3.90	3.67	3.50	3.36	3.25	3.17	3.09	3.03	2.93
25	7.77	5.57	4.68	4.18	3.86	3.63	3.46	3.32	3.21	3.13	3.05	2.99	2.89
26	7.72	5.53	4.64	4.14	3.82	3.59	3.42	3.29	3.17	3.09	3.02	2.96	2.86
27	7.68	5.49	4.60	4.11	3.79	3.56	3.39	3.26	3.14	3.06	2.98	2.93	2.83
28	7.64	5.45	4.57	4.07	3.76	3.53	3.36	3.23	3.11	3.03	2.95	2.90	2.80
29	7.60	5.42	4.54	4.04	3.73	3.50	3.33	3.20	3.08	3.00	2.92	2.87	2.77

Table 7 (Continued)

Between Groups Degrees of Freedom (Numerator)											Within Groups Degrees of Freedom
16	20	24	30	40	50	75	100	200	500	infinity	
6,169	6,208	6,234	6,258	6,286	6,302	6,323	6,334	6,352	6,361	6,366	1
99.44	99.45	99.46	99.47	99.48	99.48	99.49	99.49	99.49	99.50	99.50	2
26.83	26.69	26.60	26.50	26.41	26.35	26.27	26.23	26.18	26.14	26.12	3
14.15	14.02	13.93	13.83	13.74	13.69	13.61	13.57	13.52	13.48	13.46	4
9.68	9.55	9.47	9.38	9.29	9.24	9.17	9.13	9.07	9.04	9.02	5
7.52	7.39	7.31	7.23	7.14	7.09	7.02	6.99	6.94	6.90	6.88	6
6.27	6.15	6.07	5.98	5.90	5.85	5.78	5.75	5.70	5.67	5.65	7
5.48	5.36	5.28	5.20	5.11	5.06	5.00	4.96	4.91	4.88	4.86	8
4.92	4.80	4.73	4.64	4.56	4.51	4.45	4.41	4.36	4.33	4.31	9
4.52	4.41	4.33	4.25	4.17	4.12	4.05	4.01	3.96	3.93	3.91	10
4.21	4.10	4.02	3.94	3.86	3.80	3.74	3.70	3.66	3.62	3.60	11
3.98	3.86	3.78	3.70	3.61	3.56	3.49	3.46	3.41	3.38	3.36	12
3.78	3.67	3.59	3.51	3.42	3.37	3.30	3.27	3.21	3.18	3.16	13
3.62	3.51	3.43	3.34	3.26	3.21	3.14	3.11	3.06	3.02	3.00	14
3.48	3.36	3.29	3.20	3.12	3.07	3.00	2.97	2.92	2.89	2.87	15
3.37	3.25	3.18	3.10	3.01	2.96	2.89	2.86	2.80	2.77	2.75	16
3.27	3.16	3.08	3.00	2.92	2.86	2.79	2.76	2.70	2.67	2.65	17
3.19	3.07	3.00	2.91	2.83	2.78	2.71	2.68	2.62	2.59	2.57	18
3.12	3.00	2.92	2.84	2.76	2.70	2.63	2.60	2.54	2.51	2.49	19
3.05	2.94	2.86	2.77	2.69	2.63	2.56	2.53	2.47	2.44	2.42	20
2.99	2.88	2.80	2.72	2.63	2.58	2.51	2.47	2.42	2.38	2.36	21
2.94	2.83	2.75	2.67	2.58	2.53	2.46	2.42	2.37	2.33	2.31	22
2.89	2.78	2.70	2.62	2.53	2.48	2.41	2.37	2.32	2.28	2.26	23
2.85	2.74	2.66	2.58	2.49	2.44	2.36	2.33	2.27	2.23	2.21	24
2.81	2.70	2.62	2.54	2.45	2.40	2.32	2.29	2.23	2.19	2.17	25
2.77	2.66	2.58	2.50	2.41	2.36	2.28	2.25	2.19	2.15	2.13	26
2.74	2.63	2.55	2.47	2.38	2.33	2.25	2.21	2.16	2.12	2.10	27
2.71	2.60	2.52	2.44	2.35	2.30	2.22	2.18	2.13	2.09	2.06	28
2.68	2.57	2.49	2.41	2.32	2.27	2.19	2.15	2.10	2.06	2.03	29

Continued

Table 7 (Continued)

Within Groups Degrees of Freedom	Between Groups Degrees of Freedom (Numerator)												
	1	2	3	4	5	6	7	8	9	10	11	12	14
30	7.56	5.39	4.51	4.02	3.70	3.47	3.30	3.17	3.06	2.98	2.90	2.84	2.74
32	7.50	5.34	4.46	3.97	3.66	3.42	3.25	3.12	3.01	2.94	2.86	2.80	2.70
34	7.44	5.29	4.42	3.93	3.61	3.38	3.21	3.08	2.97	2.89	2.82	2.76	2.66
36	7.39	5.25	4.38	3.89	3.58	3.35	3.18	3.04	2.94	2.86	2.78	2.72	2.62
38	7.35	5.21	4.34	3.86	3.54	3.32	3.15	3.02	2.91	2.82	2.75	2.69	2.59
40	7.31	5.18	4.31	3.83	3.51	3.29	3.12	2.99	2.88	2.80	2.73	2.66	2.56
42	7.27	5.15	4.29	3.80	3.49	3.26	3.10	2.96	2.86	2.77	2.70	2.64	2.54
44	7.24	5.12	4.26	3.78	3.46	3.24	3.07	2.94	2.84	2.75	2.68	2.62	2.52
46	7.21	5.10	4.24	3.76	3.44	3.22	3.05	2.92	2.82	2.73	2.66	2.60	2.50
48	7.19	5.08	4.22	3.74	3.42	3.20	3.04	2.90	2.80	2.71	2.64	2.58	2.48
50	7.17	5.06	4.20	3.72	3.41	3.18	3.02	2.88	2.78	2.70	2.62	2.56	2.46
55	7.12	5.01	4.16	3.68	3.37	3.15	2.98	2.85	2.75	2.66	2.59	2.53	2.43
60	7.08	4.98	4.13	3.65	3.34	3.12	2.95	2.82	2.72	2.63	2.56	2.50	2.40
65	7.04	4.95	4.10	3.62	3.31	3.09	2.93	2.79	2.70	2.61	2.54	2.47	2.37
70	7.01	4.92	4.08	3.60	3.29	3.07	2.91	2.77	2.67	2.59	2.51	2.45	2.35
80	6.96	4.88	4.04	3.56	3.25	3.04	2.87	2.74	2.64	2.55	2.48	2.41	2.32
100	6.90	4.82	3.98	3.51	3.20	2.99	2.82	2.69	2.59	2.51	2.43	2.36	2.26
125	6.84	4.78	3.94	3.47	3.17	2.95	2.79	2.65	2.56	2.47	2.40	2.33	2.23
150	6.81	4.75	3.91	3.44	3.14	2.92	2.76	2.62	2.53	2.44	2.37	2.30	2.20
200	6.76	4.71	3.88	3.41	3.11	2.90	2.73	2.60	2.50	2.41	2.34	2.28	2.17
400	6.70	4.66	3.83	3.36	3.06	2.85	2.69	2.55	2.46	2.37	2.29	2.23	2.12
1000	6.66	4.62	3.80	3.34	3.04	2.82	2.66	2.53	2.43	2.34	2.26	2.20	2.09
Infinity	6.64	4.60	3.78	3.32	3.02	2.80	2.64	2.51	2.41	2.32	2.24	2.18	2.07

Table 7 (Continued)

Between Groups Degrees of Freedom (Numerator)											Within Groups Degrees of
16	20	24	30	40	50	75	100	200	500	Infinity	Freedom
2.66	2.55	2.47	2.38	2.29	2.24	2.16	2.13	2.07	2.03	2.01	30
2.62	2.51	2.42	2.34	2.25	2.20	2.12	2.08	2.02	1.98	1.96	32
2.58	2.47	2.38	2.30	2.21	2.15	2.08	2.04	1.98	1.94	1.91	34
2.54	2.43	2.35	2.26	2.17	2.12	2.04	2.00	1.94	1.90	1.87	36
2.51	2.40	2.32	2.22	2.14	2.08	2.00	1.97	1.90	1.86	1.84	38
2.49	2.37	2.29	2.20	2.11	2.05	1.97	1.94	1.88	1.84	1.81	40
2.46	2.35	2.26	2.17	2.08	2.02	1.94	1.91	1.85	1.80	1.78	42
2.44	2.32	2.24	2.15	2.06	2.00	1.92	1.88	1.82	1.78	1.75	44
2.42	2.30	2.22	2.13	2.04	1.98	1.90	1.86	1.80	1.76	1.72	46
2.40	2.28	2.20	2.11	2.02	1.96	1.88	1.84	1.78	1.73	1.70	48
2.39	2.26	2.18	2.10	2.00	1.94	1.86	1.82	1.76	1.71	1.68	50
2.35	2.23	2.15	2.06	1.96	1.90	1.82	1.78	1.71	1.66	1.64	55
2.32	2.20	2.12	2.03	1.93	1.87	1.79	1.74	1.68	1.63	1.60	60
2.30	2.18	2.09	2.00	1.90	1.84	1.76	1.71	1.64	1.60	1.56	65
2.28	2.15	2.07	1.98	1.88	1.82	1.74	1.69	1.62	1.56	1.53	70
2.24	2.11	2.03	1.94	1.84	1.78	1.70	1.65	1.57	1.52	1.49	80
2.19	2.06	1.98	1.89	1.79	1.73	1.64	1.59	1.51	1.46	1.43	100
2.15	2.03	1.94	1.85	1.75	1.68	1.59	1.54	1.46	1.40	1.37	125
2.12	2.00	1.91	1.83	1.72	1.66	1.56	1.51	1.43	1.37	1.33	150
2.09	1.97	1.88	1.79	1.69	1.62	1.53	1.48	1.39	1.33	1.28	200
2.04	1.92	1.84	1.74	1.64	1.57	1.47	1.42	1.32	1.24	1.19	400
2.01	1.89	1.81	1.71	1.61	1.54	1.44	1.38	1.28	1.19	1.11	1000
1.99	1.87	1.79	1.69	1.59	1.52	1.41	1.36	1.25	1.15	1.00	Infinity

Table 8
Studentized Range Statistic (*q*) for the .05 Level

Degrees of Freedom	Number of Treatments										
	2	3	4	5	6	7	8	9	10	11	12
5	3.64	4.60	5.22	5.67	6.03	6.33	6.58	6.80	6.99	7.17	7.32
6	3.46	4.34	4.90	5.30	5.63	5.90	6.12	6.32	6.49	6.65	6.79
7	3.34	4.16	4.68	5.06	5.36	5.61	5.82	6.00	6.16	6.30	6.43
8	3.26	4.04	4.53	4.89	5.17	5.40	5.60	5.77	5.92	6.05	6.18
9	3.20	3.95	4.41	4.76	5.02	5.24	5.43	5.59	5.74	5.87	5.98
10	3.15	3.88	4.33	4.65	4.91	5.12	5.30	5.46	5.60	5.72	5.83
11	3.11	3.82	4.26	4.57	4.82	5.03	5.20	5.35	5.49	5.61	5.71
12	3.08	3.77	4.20	4.51	4.75	4.95	5.12	5.27	5.39	5.51	5.61
13	3.06	3.73	4.15	4.45	4.69	4.88	5.05	5.19	5.32	5.43	5.53
14	3.03	3.70	4.11	4.41	4.64	4.83	4.99	5.13	5.25	5.36	5.46
15	3.01	3.67	4.08	4.37	4.59	4.78	4.94	5.08	5.20	5.31	5.40
16	3.00	3.65	4.05	4.33	4.56	4.74	4.90	5.03	5.15	5.26	5.35
17	2.98	3.63	4.02	4.30	4.52	4.70	4.86	4.99	5.11	5.21	5.31
18	2.97	3.61	4.00	4.28	4.49	4.67	4.82	4.96	5.07	5.17	5.27
19	2.96	3.59	3.98	4.25	4.47	4.65	4.79	4.92	5.04	5.14	5.23
20	2.95	3.58	3.96	4.23	4.45	4.62	4.77	4.90	5.01	5.11	5.20
24	2.92	3.53	3.90	4.17	4.37	4.54	4.68	4.81	4.92	5.01	5.10
30	2.89	3.49	3.85	4.10	4.30	4.46	4.60	4.72	4.82	4.92	5.00
40	2.86	3.44	3.79	4.04	4.23	4.39	4.52	4.63	4.73	4.82	4.90
60	2.83	3.40	3.74	3.98	4.16	4.31	4.44	4.55	4.65	4.73	4.81
120	2.80	3.36	3.68	3.92	4.10	4.24	4.36	4.47	4.56	4.64	4.71
Infinity	2.77	3.31	3.63	3.86	4.03	4.17	4.29	4.39	4.47	4.55	4.62

Table 9
Studentized Range Statistic (*q*) for the .01 Level

Degrees of Freedom	Number of Treatments										
	2	3	4	5	6	7	8	9	10	11	12
5	5.70	6.98	7.80	8.42	8.91	9.32	9.67	9.97	10.24	10.48	10.70
6	5.24	6.33	7.03	7.56	7.97	8.32	8.61	8.87	9.10	9.30	9.48
7	4.95	5.92	6.54	7.01	7.37	7.68	7.94	8.17	8.37	8.55	8.71
8	4.75	5.64	6.20	6.62	6.96	7.24	7.47	7.68	7.86	8.03	8.18
9	4.60	5.43	5.96	6.35	6.66	6.91	7.13	7.33	7.49	7.65	7.78
10	4.48	5.27	5.77	6.14	6.43	6.67	6.87	7.05	7.21	7.36	7.49
11	4.39	5.15	5.62	5.97	6.25	6.48	6.67	6.84	6.99	7.13	7.25
12	4.32	5.05	5.50	5.84	6.10	6.32	6.51	6.67	6.81	6.94	7.06
13	4.26	4.96	5.40	5.73	5.98	6.19	6.37	6.53	6.67	6.79	6.90
14	4.21	4.89	5.32	5.63	5.88	6.08	6.26	6.41	6.54	6.66	6.77
15	4.17	4.84	5.25	5.56	5.80	5.99	6.16	6.31	6.44	6.55	6.66
16	4.13	4.79	5.19	5.49	5.72	5.92	6.08	6.22	6.35	6.46	6.56
17	4.10	4.74	5.14	5.43	5.66	5.85	6.01	6.15	6.27	6.38	6.48
18	4.07	4.70	5.09	5.38	5.60	5.79	5.94	6.08	6.20	6.31	6.41
19	4.05	4.67	5.05	5.33	5.55	5.73	5.89	6.02	6.14	6.25	6.34
20	4.02	4.64	5.02	5.29	5.51	5.69	5.84	5.97	6.09	6.19	6.28
24	3.96	4.55	4.91	5.17	5.37	5.54	5.69	5.81	5.92	6.02	6.11
30	3.89	4.45	4.80	5.05	5.24	5.40	5.54	5.65	5.76	5.85	5.93
40	3.82	4.37	4.70	4.93	5.11	5.26	5.39	5.50	5.60	5.69	5.76
60	3.76	4.28	4.59	4.82	4.99	5.13	5.25	5.36	5.45	5.53	5.60
120	3.70	4.20	4.50	4.71	4.87	5.01	5.12	5.21	5.30	5.37	5.44
Infinity	3.64	4.12	4.40	4.60	4.76	4.88	4.99	5.08	5.16	5.23	5.29

Table 10
Minimum Values for Significance of a Pearson r[1]

df	.05 level	.01 level	.001 level
1	.996	.999	.999
2	.950	.990	.999
3	.878	.959	.991
4	.811	.917	.974
5	.754	.874	.951
6	.707	.834	.925
7	.666	.798	.898
8	.632	.765	.872
9	.602	.735	.847
10	.576	.708	.823
11	.553	.684	.801
12	.532	.661	.780
13	.514	.641	.760
14	.497	.623	.742
15	.482	.606	.725
16	.468	.590	.708
17	.456	.575	.693
18	.444	.561	.679
19	.433	.549	.665
20	.423	.537	.652
25	.381	.487	.597
30	.349	.449	.554
35	.325	.418	.519

df	.05 level	.01 level	.001 level
40	.304	.393	.490
45	.288	.372	.465
50	.273	.354	.443
60	.250	.325	.408
70	.232	.302	.380
80	.217	.283	.357
90	.205	.267	.338
100	.195	.254	.321

[1]Abridged from Fisher, R. A., & Yates, F. *Statistical tables for biological, agricultural, and medical research.* Edinburgh: Oliver Boyd Ltd.

Table 11
Critical Values of Chi Square[1]

df	.05 level	.01 level	.001 level	df	.05 level	.01 level	.001 level
1	3.841	6.635	10.827	24	36.415	42.980	51.179
2	5.991	9.210	13.815	25	37.652	44.314	52.620
3	7.815	11.345	16.268	26	38.885	45.642	54.052
4	9.488	13.277	18.465	27	40.113	46.963	55.476
5	11.070	15.086	20.517	28	41.337	48.278	56.893
6	12.592	16.812	22.457	29	42.557	49.588	58.302
7	14.067	18.475	24.322	30	43.773	50.892	59.703
8	15.507	20.090	26.125				
9	16.919	21.666	27.877				
10	18.307	23.209	29.588				
11	19.675	24.725	31.264				
12	21.026	26.217	32.909				
13	22.362	27.688	34.528				
14	23.685	29.141	36.123				
15	24.996	30.578	37.697				
16	26.296	32.000	39.252				
17	27.587	33.409	40.790				
18	28.869	34.805	42.312				
19	30.144	36.191	43.820				
20	31.410	37.566	45.315				
21	32.671	38.932	46.797				
22	33.924	40.289	48.268				
23	35.172	41.638	49.728				

[1]Abridged from Fisher, R. A., & Yates, F. *Statistical tables for biological, agricultural, and medical research*. Edinburgh: Oliver Boyd Ltd.

Table 12
Critical Values of U for the .05 Level[1]

N_1	N_2											
	9	10	11	12	13	14	15	16	17	18	19	20
1												
2	0	0	0	1	1	1	1	1	2	2	2	2
3	2	3	3	4	4	5	5	6	6	7	7	8
4	4	5	6	7	8	9	10	11	11	12	13	13
5	7	8	9	11	12	13	14	15	17	18	19	20
6	10	11	13	14	16	17	19	21	22	24	25	27
7	12	14	16	18	20	22	24	26	28	30	32	34
8	15	17	19	22	24	26	29	31	34	36	38	41
9	17	20	23	26	28	31	34	37	39	42	45	48
10	20	23	26	29	33	36	39	42	45	48	52	55
11	23	26	30	33	37	40	44	47	51	55	58	62
12	26	29	33	37	41	45	49	53	57	61	65	69
13	28	33	37	41	45	50	54	59	63	67	72	76
14	31	36	40	45	50	55	59	64	67	74	78	83
15	34	39	44	49	54	59	64	70	75	80	85	90
16	37	42	47	53	59	64	70	75	81	86	92	98
17	39	45	51	57	63	67	75	81	87	93	99	105
18	42	48	55	61	67	74	80	86	93	99	106	112
19	45	52	58	65	72	78	85	92	99	106	113	119
20	48	55	62	69	76	83	90	98	105	112	119	127

[1]Abridged from Wilcoxon, F. (1949). *Some rapid approximate statistical procedures.* New York: American Cyanamid Co.

Table 13
Critical Values of *U* for the .01 Level[1]

N_1	N_2 9	10	11	12	13	14	15	16	17	18	19	20
1												
2											0	0
3	0	0	0	1	1	1	2	2	2	2	3	3
4	1	2	2	3	3	4	5	5	6	6	7	8
5	3	4	5	6	7	7	8	9	10	11	12	13
6	5	6	7	9	10	11	12	13	15	16	17	18
7	7	9	10	12	13	15	16	18	19	21	22	24
8	9	11	13	15	17	18	20	22	24	26	28	30
9	11	13	16	18	20	22	24	27	29	31	33	36
10	13	16	18	21	24	26	29	31	34	37	39	42
11	16	18	21	24	27	30	33	36	39	42	45	48
12	18	21	24	27	31	34	37	41	44	47	51	54
13	20	24	27	31	34	38	42	45	49	53	56	60
14	22	26	30	34	38	42	46	50	54	58	63	67
15	24	29	33	37	42	46	51	55	60	64	69	73
16	27	31	36	41	45	50	55	60	65	70	74	79
17	29	34	39	44	49	54	60	65	70	75	81	86
18	31	37	42	47	53	58	64	70	75	81	87	92
19	33	39	45	51	56	63	69	74	81	87	93	99
20	36	42	48	54	60	67	73	79	86	92	99	105

[1]Abridged from Wilcoxon, F. (1949). *Some rapid approximate statistical procedures.* New York: American Cyanamid Co.

Table 14
Critical Values of *T* for Wilcoxon's Matched-Pairs Test[1]

df	.05 level	.01 level
6	0	—
7	2	—
8	4	0
9	6	2
10	8	3
11	11	5
12	14	7
13	17	10
14	21	13
15	25	16
16	30	20
17	35	23
18	40	28
19	46	32
20	52	38
21	59	43
22	66	49
23	73	55
24	81	61
25	89	68

[1]From Wilcoxon, F. (1949). *Some rapid approximate statistical procedures*. New York: American Cyanamid Co.

Index

Notes:

Notes:

Notes:

Notes:

Notes: